W9-BCG-413

EXPLAINING
SCIENCE

Science and Its Conceptual Foundations
David L. Hull, Editor

Ronald N. Giere

EXPLAINING

SCIENCE

A COGNITIVE APPROACH

The University of Chicago Press
Chicago and London

SIENA COLLEGE LIBRARY

The University of Chicago Press, Chicago 60637
The University of Chicago Press, Ltd., London

© 1988 by The University of Chicago
All rights reserved. Published 1988
Paperback edition 1990
Printed in the United States of America

97 96 95 94 93 92 91 90   5 4 3 2

Library of Congress Cataloging in Publication Data

Giere, Ronald N.
  Explaining science: a cognitive approach / Ronald N. Giere.
      p.   cm. — (Science and its conceptual foundations series)
  Bibliography: p.
  Includes index.
  ISBN 0-226-29205-3 (cloth)
  ISBN 0-226-29206-1 (paper)
  1. Science—Philosophy.   2. Science—Social aspects.
3. Cognition.   4. Realism.   I. Title.   II. Series: Science and its
conceptual foundations.
Q175.G4889   1988
500—dc19   87-27033
                                                        CIP

This book is printed on acid-free paper.

Q
175
.G4889
1970

For Barbara

Contents

List of Figures		xi
Preface		xv
Acknowledgments		xix
1	**Toward a Unified Cognitive Theory of Science**	**1**
	What Might a Cognitive Theory of Science Be?	1
	Rationality, Relativism, and Cognition	2
	Representation and Judgment	5
	Naturalistic Realism	7
	Can the Philosophy of Science Be Naturalized?	8
	Must the Naturalistic Study of Science Be Viciously Circular?	10
	Evolutionary Naturalism	12
	What Might a Cognitive Theory of Science Be Like?	16
	A Role for History?	18
	Overview of This Book	19
2	**Theories of Science**	**22**
	Logical Empiricism	22
	The Social Structure of Science	29
	Paradigms and Revolutions	32
	Programs and Traditions	38
	Constructive Empiricism	46
	The Strong Program	50
	Laboratory Studies	56
	The Sociological Analysis of Scientists' Discourse	59
3	**Models and Theories**	**62**
	The Science Textbook	62

The Organization of a Mechanics Text   64
The Linear Oscillator   68
Interpretation and Identification   74
The Laws of Motion   76
Models and Hypotheses   78
What Is a Scientific Theory?   82
What about Axiomatic Presentations of Mechanics?   87
Beyond Classical Mechanics   89

4   Constructive Realism   **92**
Respects of Similarity   92
Varieties of Empiricism   94
Unrestricted Realism   96
Metaphysical Realism   98
Modal Realism   98
Laws as Universal Generalizations   102
Causal Models and Causal Explanations   104
Realistic Rejoinders   106

5   Realism in the Laboratory   **111**
Contingency and Negotiation   112
Producing Protons   115
Using Protons   120
Experimentation and Realism   124
The Limitations of Empiricism   128
The Limitations of Constructivism   130
Geometrical Cognition in Nuclear Research   133
The Role of Technology in Scientific Research   136

6   Scientific Judgment   **141**
Scientists as Decision Makers   141
Basic Decision Models   143
Bayesian Decision Models   145
Are Scientists Bayesian Agents?   149
Satisficing Models   157
Scientists as Satisficers   161
Experimental Tests   165
Philosophical Objections   168
The Role of Probability in Science   172

7 Models and Experiments **179**
Models of the Nuclear Potential 180
Background to the Pursuit of Relativistic Dirac Models 184
Response to the New Data 191
Why Successful Predictions Matter 198
Further Evidence 203
The Design and Execution of an Experimental Test 208
Cognitive Resources and Scientific Interests 213
An Evolutionary Picture 222
The Future of Dirac Models in Nuclear Physics 222

8 Explaining the Revolution in Geology **227**
Contractionist Models 228
Wegener and Continental Drift 229
Wegener's Critics 234
Oceanography and Paleomagnetism 241
Seafloor Spreading 249
The Vine-Matthews Hypothesis 252
The Juan de Fuca Ridge 257
The Vindication of Seafloor Spreading 261
Mobilism Becomes a Satisfactory Option 270
Revolution or Evolution? 275

Epilogue: Reflexive Reflections **279**

Notes 281
References 295
Index 315

Figures

1.1 Four possible combinations of views generated by a double dichotomy along the dimensions of representation and judgment 8

2.1 A logical empiricist picture of a scientific theory 25

2.2 Laudan's reticulated model 45

3.1 Contents page of a typical mechanics text 67

3.2 The mass and spring system used to illustrate Hooke's law and simple harmonic motion 69

3.3 Position and velocity as functions of time for the mass and spring system of figure 3.2 69

3.4 A simple pendulum 71

3.5 The state of a simple harmonic oscillator in position-momentum space 72

3.6 The state of a simple harmonic oscillator in position-momentum-time space 72

3.7 The position of a damped harmonic oscillator as a function of time 74

3.8 Relationships among sets of statements, models, and real systems 83

3.9 A partial representation of the families of models associated with classical mechanics 83

3.10 A picture of a theory, showing a family of models and related real systems 85

5.1 The layout of the Indiana University Cyclotron Facility 116

5.2 A cutaway view of the main stage cyclotron as seen from above 118

5.3 A representation of the nuclear reaction $p-C^{14}-N^{14}-n$. 121

5.4 An experimental setup for (p,n) experiments at IUCF 121

5.5 An experimental setup for rat navigation experiments   134
6.1 The basic structure of a decision   144
6.2 A representation of a scientist's decision problem in
       choosing between two models   147
6.3 The responses of a single subject to random samples from
       a binomial population with a 70–30 success-to-failure
       ratio   151
6.4 A representation of a diagnostic test with false positives of
       5 percent   154
6.5 A decision problem for an "open-minded" satisficer
       choosing between two models   164
6.6 A schematic representation of an experimental test, with
       disjoint ranges of possible results for two different
       models of the system of interest   166
6.7 Two drawings employed in research on students' use of
       models in thinking about mechanical systems   175
6.8 Wason's selection task   177
7.1 A schematic representation of Rutherford's experiment   181
7.2 A Schroedinger equation and potentials for the elastic
       scattering of 200 MeV protons by calcium 40   183
7.3 A schematic representation of the experimental setup for
       measuring differential cross-sections   186
7.4 A schematic representation of the experimental setup for
       measuring polarization   186
7.5 The first data on polarization measured at 800 MeV for a
       calcium 40 target   187
7.6 A schematic representation of the experimental setup for
       measuring spin rotation   188
7.7 The first data on spin rotation measured at 500 MeV for a
       calcium 40 target   189
7.8 A Dirac equation with a graph of the two most important
       relativistic potentials   193
7.9 The decision matrix for a choice between Schroedinger
       and Dirac models   194
7.10 A comparison of Clark's predicted curve for Q with the
       existing data   196
7.11 A comparison of all three types of data with both a
       standard impulse approximation and with the
       relativistic impulse approximation   207
7.12 Plots of the spin rotation parameter for one Schroedinger
       model and one Dirac model   209

7.13 A comparison of the initial 200 MeV spin rotation data
          with one Schroedinger model and one Dirac model   211

7.14 The final comparison of 200 MeV spin rotation data with
          two Schroedinger models and one Dirac model   221

8.1 The structure of the decision problem faced by Wegener
          and later by others who seriously considered the
          possibility of choosing a mobilist approach   231

8.2 The worldwide system of midocean ridges as it was
          conceived around 1960   243

8.3 A profile of the Mid-Atlantic Ridge   243

8.4 An anomaly map of the magnetic field along the ocean
          floor off the coast of western North America   244

8.5 The gradual refinement of time scales for geomagnetic
          reversals based on potassium-argon dating techniques
          applied to terrestrial rocks   245

8.6 The apparent position of the magnetic north pole for
          several continents as a function of time   247

8.7 Holmes's model of convection currents splitting a
          continent and producing a new ocean   250

8.8 Hess's model of convection currents and seafloor spreading 252

8.9 Vine and Matthews's comparison of the magnetic data
          across the Carlsberg Ridge   255

8.10 Computed magnetic profiles for several models of ocean
          ridges   256

8.11 Wilson's reconstruction of the San Andreas Fault   259

8.12 Vine and Wilson's comparison of the magnetic profiles of
          the East Pacific Rise and the Juan de Fuca Ridge   260

8.13 The magnetic profile of the Reykjanes Ridge produced by
          Heirtzler   263

8.14 The *Eltanin*-19 magnetic profile of the Pacific-Antarctic
          Ridge   266

8.15 A comparison of magnetic reversals revealed in deep-sea
          sediments from several locations in the south Atlantic   269

8.16 Vine's comparison of seafloor spreading rates for three
          ridges   271

Preface

I have set out the aims and general conclusions of this book in the very first chapter. The last section contains a brief summary of the whole work. Rather than attempting a similar summary here, I will use this opportunity to explain, in more personal terms, how I came to my present point of view.

I began graduate work in philosophy after having completed a graduate degree in physics. The year, 1962, is significant because it marked the appearance of Thomas Kuhn's now-famous book on scientific revolutions. I read Kuhn with great excitement because his words rang true to my recent experience as an aspiring physicist. But from a philosophical point of view it seemed that Kuhn had merely rediscovered Hume's problem of induction in historical garb and had succumbed to a "psychologistic" position that led straight to epistemological relativism. That was philosophically unacceptable.

I began dissertation research in the area then known as "probability and induction." I was troubled, however, because philosophical discussions of induction seemed so remote from scientific practice as I had known it. Much of the remoteness arose because the philosophical idiom derived from mathematical logic, whereas scientists speak in natural languages enriched by mathematics. By good fortune I was led to works in statistics by both statisticians and social scientists. There I found versions of standard philosophical problems in a more scientific idiom. I began to think of the philosophical area no longer as "probability and induction" but as "the foundations of probability and statistical inference" on the model of "the foundations of physics."

This program, however, contained internal conflicts that, while always recognized, only slowly came to occupy the center of my attention. I found I was rejecting major tenets of the philosophical tradition regarding, for example, the nature of probability, the possibility of an inductive logic, and the existence of an epistemological foundation for scientific knowledge. Yet I continued to pursue the general program of elucidating "the foundations of scientific inference," substituting the language of statistical theory for that of mathematical logic. Again I explicitly rejected the option of pursuing research in the foundations of physics because to do it well seemed to me to require the talents of a good theoretical physicist. Had I thought I had such talents, I would not have left physics. Yet I did pursue research in the foundations of probability and statistics even though I had no formal training in either probability theory or mathematical statistics.

As these internal conflicts became more pressing, I began to lose my faith in the general program, even in the rationale for pursuing any such program. My skepticism progressed to the point that I now believe there are no special philosophical foundations to any science. There is only deep theory, which, however, is part of science itself. And there are no special philosophical methods for plumbing the theoretical depths of any science. There are only the methods of the sciences themselves. Moreover, the people best equipped to engage in such pursuits are not those trained as philosophers, but those totally immersed in the scientific subject matter— namely, scientists.

In my studies of statistical theory I learned several things that were to prove useful later. Most important was to think of scientific reasoning not so much as a process of inference but as one of decision making. And among decision strategies I was impressed most with the virtues of "satisficing" over maximization. Other topics I had pursued as sidelines also proved useful. One was the view of scientific theories not as empirical statements but as definitions of models variously related to the real world. Another was biology, particularly evolutionary theory. I slowly came to appreciate the power of selectionist models as opposed to the more straightforwardly causal models of physics. Finally, I had a long-standing general concern with possible connections between the philosophy of science and related areas, particularly the history of science.

The turning point came about 1982. I began reading some recent works in the sociology of science, particularly several based on the study of scientists in a laboratory setting. I was attracted by the idea of investigating how scientists actually do science, while at the same time I was repelled by the conclusion that science is purely a social construct. But having given up the idea that science embodies some special form of rationality that philosophers might uncover, I was unsure how to formulate my objections.

The resolution once again came by a fortuitous encounter with other disciplines, this time the cognitive sciences. There one finds models of cognitive agents who develop representations of the world and make judgments about both the world and their representations of it. It took no great leap of imagination to think of scientists as cognitive agents and of scientific models as a special type of representation. Likewise, scientists' decision making about models can easily be seen as an exercise of ordinary human judgment. I now had the ingredients for a view of science that combined the features I sought.

First, the view is thoroughly naturalistic, requiring no special type of rationality beyond the effective use of available means to achieve desired goals. Second, there is room for a modest yet robust scientific realism that insists scientists are at least sometimes successful in their attempts to represent the causal structure of the world. Third, the view allows scientists to be real people with a full range of human interests while also being cognitive agents engaged in something like "the pursuit of truth." Finally, the view makes possible an account of scientific development as a natural evolutionary process.

From some traditional philosophical perspectives, a naturalized philosophy of science is not a part of philosophy at all. If one regards the philosophy of science as a part of epistemology, for example, the dominant enterprise has been to show how claims of scientific knowledge can be justified in some suitably noncircular manner. The philosophical goal, in short, has been to provide some extrascientific foundation for scientific claims. This project now seems to me merely a modern, secular version of the medieval project of providing philosophical proofs of the existence of God. Such a project might better be labeled "the theology of science" or "scientific apologetics."

There is another philosophical tradition of equal antiquity. This tradition views philosophy simply as the search for a general understanding of the world—including the activities of human beings. It is this tradition that over the centuries gave birth to the special sciences that now dominate the intellectual world. Perhaps a naturalized philosophy of science is destined to become part of a new special science—a science of science or, more popularly, science and technology studies. But these are not yet disciplines. Until the time such disciplines become established, philosophy provides a haven for theoretical studies of science as a human activity. Given the importance of science in the modern world, the need for such studies is undeniable. The label is not significant. What matters is that philosophers of science play a useful cultural role.

Acknowledgments

In completing this work I have incurred many debts to individuals, to institutions, and to publishers. I can only hope that those I have forgotten will be forgiving.

I owe long-overdue thanks to Max Black for encouraging the philosophical development of a refugee from physics. Wesley Salmon, who was my senior colleague for a half-dozen years, provided a model of philosophical integrity and clarity that I much respect. Others who have provided much-appreciated guidance and professional opportunities include Adolf Grunbaum, Ernan McMullin, Dudley Shapere, Nick Rescher, Abner Shimony, and Pat Suppes. I deeply regret the untimely death of the statistician Alan Birnbaum, who was my host at the Courant Institute in the academic year 1971–72.

Over the years I have benefited greatly from ongoing discussions with Bas van Fraassen, who occupied the office next to mine in 1968. It was from Bas that I first learned the power of a model-theoretic approach to the philosophy of science. His patience in refuting my arguments for realism never faltered, although it was never rewarded. I argued less with Fred Suppe because we have been in much closer philosophical agreement all along. I have also learned much from many years of disagreeing with Larry Laudan. I thank him and Rachel Laudan for having arranged for me a year as a senior visiting fellow at the Center for Philosophy of Science in Pittsburgh. My fellow fellows, George Gale, Ron Laymon, Tom Nickles, Jim Woodward, and John Worrall, kindly tolerated my endless disquisitions on the virtues of a naturalistic approach to the philosophy of science. Hank Frankel first led me to appreciate the value of geology as a case study for the philosophy of science, and he continues to give much-needed assistance in this area. Paul and Patricia Churchland sparked my interest in the cognitive sciences and furnished the occasion of my first meeting with Don Campbell, who has been a valued source of inspiration and encouragement. David Hull and Michael Ruse have also kindly supported my interests in evolutionary models of science. In addition they have generously provided guidance and support for my students John Beatty and Ken Waters. Arthur Fine and Paul Teller also lent encouragement when I most needed it.

I should also thank a number of other people in the psychology and sociology of science for being so kind to a philosophical interloper: David Bloor, Werner Callebaut, Harry Collins, Howard Gruber, Karin Knorr-Cetina, Bruno Latour, Trevor Pinch, Ryan Tweney, and Steve Woolgar.

Among former colleagues at Indiana University, I am deeply indebted to Chuck Foster and Ed Stephenson for providing my entrée into the world of contemporary nuclear physics. Without their help this would be a much less interesting book. Tom Gieryn, David Smith, and John Woodcock, the other members of a faculty seminar on science and technology studies, did much to brighten my spirits over the past several years. Finally, particular thanks go to Sam Preus, a colleague in religious studies, for encouraging me to borrow from the title of his book on explaining religion.

Turning to institutions, I owe my greatest debt to the National Science Foundation's Program on History and Philosophy of Science for support of this and earlier projects. Special thanks are due the director of that program, Ronald Overmann. Other support was provided by the Center for Philosophy of Science at the University of Pittsburgh and the Office of Research and Graduate Development and the Institute for Advanced Study at Indiana University.

Finally, I thank the following publishers for permission to use material from several of my recent articles:

The University of Notre Dame Press, for material from "Toward a Unified Theory of Science," in *Science and Reality,* ed. J. T. Cushing, C. F. Delaney, and Gary Gutting (Notre Dame, Ind., 1984), pp. 5–31.

The University of Chicago Press, for material from "Constructive Realism," in *Images of Science,* ed. P. M. Churchland and C. A. Hooker (Chicago, 1985), pp. 75–98.

The Philosophy of Science Association, for material from "Philosophy of Science Naturalized," in *Philosophy of Science* 52 (1985): 331–56; and from "Background Knowledge in Science: A Naturalistic Critique," in *PSA 1984,* vol. 2, ed. P. D. Asquith and P. Kitcher (East Lansing, Mich., 1986), pp. 664–71.

Martinus Nijhoff, for material from "The Cognitive Study of Science," in *The Process of Science: Contemporary Philosophical Approaches to Understanding Science,* ed. N. Nersessian (Dordrecht, 1987), pp. 139–59.

The University Press of America, for material from "Toward a Cognitive Theory of Science," in *Scientific Inquiry in Philosophical Perspective,* ed. N. Rescher (Lanham, Md., 1987), pp. 117–50.

1

Toward a Unified Cognitive Theory of Science

What Might a Cognitive Theory of Science Be?

Science, together with science-based technology, has been a growing force in Western culture for three centuries. Its influence is now felt in virtually every society throughout the world. Among the forces operative in contemporary cultures, science ranks with government, the military, commerce, and religion.

One need say no more to justify the study of science as a cultural phenomenon. But what do I mean by a *theory* of science, and why do I call it *cognitive?*

WHY A *THEORY* OF SCIENCE?

'Theory' is a vague and often primarily honorific term that shows up in such diverse linguistic contexts as "literary theory," "critical theory," "the theory of justice," "number theory," and "the atomic theory of matter." When I speak of a theory of science, I mean 'theory' in the sense of a *scientific* theory. A theory of science would thus serve to explain the phenomenon of science itself in roughly the way that scientific theories explain other natural phenomena. My view, therefore, is that the study of science as a cultural enterprise is itself a *science*. To be sure, it is a *human* science, and that raises questions about the extent to which any human science can be like such sciences as physics or biology. I shall not be diverted by such general questions here.[1]

WHY *COGNITIVE* AND WHY *UNIFIED?*

Science is a cognitive activity, which is to say it is concerned with the generation of *knowledge*. Indeed, science is now the major paradigm of a knowledge-producing enterprise. Today, one of the potentially most

1

powerful resources for studying any cognitive activity is the cluster of disciplines loosely grouped under the rubric *cognitive science*. In calling my account "a cognitive theory of science," I therefore intend it to be a broadly scientific account employing the resources of the cognitive sciences.[2]

Here one must be careful not to take too narrow a view of what constitutes cognitive science. Some equate it with cognitive psychology. Others would restrict it to the use of computational models of cognition, whether human or otherwise. I would insist on a broad view that includes parts of logic and philosophy and then runs all the way from cognitive neurobiology, through cognitive psychology and artificial intelligence, to linguistics, and on to cognitive sociology and anthropology. That is what I mean by saying that a cognitive theory of science should be *unified*. One should not put a priori restrictions on what might prove useful in explaining the phenomenon of modern science.[3]

In advocating a cognitive approach to the study of science as a human activity, I am obviously implying that there are important shortcomings to existing accounts of the nature of science. Those accounts are to be found primarily in the writings of philosophers and sociologists of science. I will comment in greater detail on some of those accounts in the next chapter. Here I will paint the contrasts with broadest brush so that the outlines of my own position will be clear from the start.

Rationality, Relativism, and Cognition

Philosophical theories of science are generally theories of scientific rationality. The scientist of philosophical theory is an ideal type, the ideally rational scientist. The actions of real scientists, when they are considered at all, are measured and evaluated by how well they fulfill the ideal. The context of science, whether personal, social, or more broadly cultural, is therefore typically regarded as irrelevant to a proper philosophical understanding of science.

Recent sociology of science, by contrast, emphasizes context above all else. Scientific knowledge, it is claimed, is totally relative to context. It is "contingent." Science, as one sociologist recently claimed, is "one hundred percent a social construct."[4]

Of course, both philosophers and sociologists have arguments they use to support and defend their respective views. Later I will consider some of those arguments. Here I will only indicate the views and explain why I regard them as unsatisfactory. These comments will provide some background for my subsequent claims that a cognitive approach is more promising. The rest of this book will be, to borrow Darwin's famous phrase, "one long argument" for these claims.

THE RATIONALITY OF SCIENCE

This subheading, which repeats the title of a recent survey of the philosophy of science (Newton-Smith 1981), well expresses the philosopher's concern with rationality. The relationship between theory and data, for example, is said to be a "rational," or even "logical," relationship. For any given theory and any given set of data, there is said to be a "rationally correct" conclusion about the extent to which those data "rationally support" the theory. The philosopher's task has been seen as one of making explicit the principles that scientists are deemed to employ intuitively in evaluating theories and then to show that these principles are indeed rationally correct.

The major philosophical problem with this approach has always been to demonstrate that a particular principle captures a relationship that is uniquely rational. In fact, addressing that problem is what most of the literature is all about. The inconclusiveness of the literature, after many years of effort, is generally taken to indicate the difficulty of the problem. But it may also be taken as a basis for suspecting that there is something fundamentally mistaken about the whole enterprise.

There is also a severe empirical problem with the typical philosopher's picture of relationships between theory and data. If scientific judgment were indeed guided by intuited principles of rationality, one would expect far more agreement among scientists than in fact exists. Of course, if one's picture of science is gathered mainly from textbooks, widespread agreement seems to be the order of the day. And it is true that there are large areas of agreement that form the background for current research. But if one looks at any area of *active* research, one almost invariably finds individuals and groups with widely divergent, and often passionately defended, opinions. Moreover, the range of opinion is not randomly distributed among scientists but typically clustered in identifiable "schools." This is the normal state of affairs at the research front.

If there are such things as principles of scientific rationality, the widespread existence of disagreeing groups in science must be the result of irrational forces or interests. One ends up with a picture of scientific research as a highly irrational activity. But most philosophers begin with the assumption that science is rationality writ large. Obviously, something is wrong.

THE SOCIAL CONSTRUCTION OF SCIENTIFIC KNOWLEDGE

Versions of this subheading now appear regularly on the covers of books in the sociology, or the sociological history, of science (Latour and Woolgar 1979; MacKenzie 1981). Science, like the law, is pictured as a thoroughly social construct. Experimental data, in this view, are just one resource among many used in social negotiations over what the content of acceptable theory will be. So are the traditional scientific virtues like simplicity. In place of the philosopher's principles of rationality one finds only the clash of competing social and professional interests.

The philosopher's charge that any such view leads to relativism is welcomed with open arms. Our scientific beliefs about the world are held to be no different in principle from Azande beliefs about witches. There is said to be no basis other than ethnocentric prejudice for our claims that we are right and the Azande wrong. Indeed, the science of the paranormal could, in different social circumstances, be normal science (Collins and Pinch 1982).

The sociological picture of science at least has the virtue of explaining the almost universal existence of disagreement at the research frontier. Disagreement in science is as natural as disagreement in the halls of Parliament; in this view, the nature of local disagreements, as well as of background agreement, is fundamentally the same in both science and politics.

The main trouble with this approach to understanding science is that it utterly fails to explain the obvious success of science, and particularly the success of science-based technology, since the seventeenth century. It makes the success of science and technology similar to the success of liberal democracy. In both cases the success would be explained primarily in terms of changing social relationships.

What such an account fails to explain is how we came to be able to produce insulin in a laboratory or to send instrument-packed rockets to photograph Uranus. Such feats require substantial social organization, to be sure, but no amount of social organizing could have produced those results even as recently as fifty years ago. There must be more going on than changing social relationships.[5]

There is indeed something important missing from the sociological account. That something is not rationality, but causal interaction between scientists and the world.

COGNITION AND REALITY

Again I have taken my subheading from a recent book, this time in cognitive psychology (Neisser 1976). The starting point for cognitive psychology, and the cognitive sciences generally, is that humans have various biologically based cognitive capacities including perception, motor control, memory, imagination, and language. People employ these capacities in everyday interactions with the world. A cognitive theory of science would attempt to explain how scientists use these capacities for interacting with the world as they go about the business of constructing modern science.

Until very recently there has been little study of scientists themselves from a cognitive perspective. As yet no standard cognitive views describe, for example, how data and scientific theory are related. This is a drawback, but it is also a blessing. The shortcomings of both the philosophy and the sociology of science stem from the fact that, in being true to their disciplinary backgrounds, both philosophers and sociologists of science have failed their subject matter—science.

Philosophers since the time of Aristotle have been custodians of the concept of rationality. They do not seriously consider the possibility that rationality might not be an especially useful concept for understanding modern science. Similarly, recent sociologists of science have been preoccupied with the idea of building a genuinely social theory of science. They have not investigated nonsocial factors. Having fewer (or at least less focused) disciplinary commitments, a cognitive theory of science is freer to be true to its subject.

Representation and Judgment

For over fifty years philosophers of science have labored to elucidate the nature of scientific theories and the criteria for choosing one theory over others. In that time they have proposed several very different sorts of answers to those two questions, but the issues have nonetheless remained important. Recent sociologists of science have concerned themselves less with the nature of theories as such, but they have devoted much effort to studying the processes by which theories come to be accepted by a majority of a scientific community.

The cognitive sciences do not provide immediate and direct answers to these questions. They do, however, suggest a general orientation and some specific concepts that provide a potentially fruitful background for considering the issues. In the framework of the cognitive sciences, theories would be some sort of representation, and the selection of a particular theory as the best available would be a matter of individual judgment.

REPRESENTATION

That humans (and animals) create internal representations of their environment (as well as of themselves) is probably the central notion in the cognitive sciences. It is the appeal to internal, mental representations, for example, that fundamentally distinguishes cognitive psychology from behaviorism. Depending on the particular field within the cognitive sciences, one finds talk of such things as "schemata," "cognitive maps," "mental models," or "frames."[6]

Later I will argue that scientific theories should be regarded as similar to the more ordinary sorts of representations studied by the cognitive sciences. There are differences to be sure. Scientific theories are more often described using written words or mathematical symbols than are the mental models of the lay person. But fundamentally the two are the same sort of thing.

Here the only feature of representations I wish to remark on is that they are just that—representations. To put it baldly, they are "internal maps" of the external world. Thus, employing the cognitive scientist's framework to think about scientific theories automatically makes one some kind of "realist." But that framework need not commit one to the hard kinds of realism criticized by both philosophers and sociologists.

Schemata are not generally described as being true or false, but as fitting the world in limited respects and degrees, and for various purposes. Cognitive scientists are not "metaphysical realists" (Putnam 1978, 1983). By the same token, there is no suggestion that the world simply presents itself without considerable internal processing or "interpretation." The mind (or brain) is not a mirror to nature (Rorty 1979).

JUDGMENT

Philosophers picture scientists as evaluating theories by appeal to some vaguely perceived principles of rational evaluation. Sociologists see a process of social negotiation based on interests and contingent needs. Cognitive scientists are not too different from sociologists. They put more emphasis than sociologists do on particular judgments by individuals, and they regard social factors as being filtered through individuals. For both, however, judgment is simply a natural activity of human beings.

A major difference between cognitive scientists and sociologists of science is that the former are more willing to investigate the effectiveness of an individual's judgmental strategies. They are more inclined themselves to judge some strategies as ineffective, or even "biased" (Nisbett and Ross 1980; Kahneman, Slovic, and Tversky 1982; Faust 1984). Sociologists are reluctant make such judgments. But "ineffective" or "biased" is not the same as the philosopher's notion of "irrational."

CATEGORICAL VERSUS HYPOTHETICAL RATIONALITY

When philosophers talk about rationality, they generally mean categorical rationality. Thus, Aristotle is reputed to have *defined* humans as rational animals. Here rationality is viewed as a property, indeed an essential property, of being human. It unambiguously distinguishes humans from other animals. And it does not come in degrees. An entity either has it or not.

There is another sense of "rational," which simply means using a known, effective means to a desired goal. This is hypothetical, or instrumental, rationality, and it does come in degrees. When cognitive scientists investigate peoples' judgmental strategies, they are at most evaluating the instrumental rationality of their subjects. What strategies are subjects using in pursuing their goals? Are they doing as well in achieving their goals as they could in the circumstances? And if not, why not? There need be no suggestion that subjects might be "irrational" in a categorical sense.

Naturalistic Realism

The philosophical fortunes of scientific realism have risen and fallen with dramatic rapidity during the past twenty years. In 1965 it had few defenders. By 1975 it was accepted by many philosophers of science. By 1985 it was again on the defensive. Yet there has been some progress, if only because the range of alternative positions has been extended and clarified (Leplin 1984; Churchland and Hooker 1985). I will discuss some of these developments in later chapters.

Here I will simply characterize scientific realism as the view that when a scientific theory is accepted, most elements of the theory are taken as representing (in some respects and to some degree) aspects of the world. Antirealism is the view that theories are accepted for some nonrepresentational virtue, such as "problem-solving effectiveness" (Laudan 1977), or for very limited representational virtues, such as saving the observable phenomena (van Fraassen 1980).

I will use the label rationalism for the view that there are rational principles for the evaluation of theories. I realize that this terminology has the misleading consequence that many "empiricists" turn out to be "rationalists," but I can think of no better term. Naturalism, by contrast, is the view that theories come to be accepted (or not) through a natural process involving both individual judgment and social interaction. No appeal to supposed rational principles of theory choice is involved.

REPRESENTATION

JUDGMENT		REALIST	ANTI-REALIST
	RATIONAL		
	NATURAL		

Figure 1.1 Four possible combinations of views generated by a double dichotomy along the dimensions of representation and judgment.

We have, then, a double dichotomy along the dimensions of representation and judgment. The result is four possible combinations of views pictured in the matrix of figure 1.1. It is interesting to see how various students of the scientific enterprise might fit into this framework.

Philosophers of science split over realism about theories, but they tend strongly to be rationalists about theory choice. Thus, Popper (1972, 1983) is a realist while Laudan (1981a, 1984a) is an anti-realist, but both are concerned to exhibit the principles of rationality underlying the choice of theories. Recent sociologists of science are almost unanimously both anti-realists about theories and naturalists about theory choice. Partly this is because sociologists tend not to distinguish the two issues of representation and judgment. In much of the sociological literature, 'rationalism' and 'realism' are synonymous, making 'naturalistic realism' a contradiction in terms.[7]

The fourth space, naturalistic realism, is the obvious place to house a cognitive theory of science, which is not to say that the space does not already contain some occupants (Campbell 1966), including some prominent philosophers of science (Boyd 1981; Churchland 1979; Gale 1984; Hooker 1978, 1987). But the location has yet to achieve the prominence it deserves. Locating the cognitive study of science in this space will, I hope, help to raise the respectability of the neighborhood.

Can the Philosophy of Science Be Naturalized?

Naturalistic realism is part of a more general naturalism—the view that all human activities are to be understood as entirely natural phenomena, as are the activities of chemicals or animals. It is also part of a long tradition of using science in the attempt to understand science itself. Like all such general viewpoints, naturalism and the scientific study of science have been "refuted" by philosophers dozens of times. What makes naturalism attractive now, however, is not that such philosophical arguments have themselves been "refuted" but that the cognitive sciences have become increasingly successful empirically. This is the historical pattern. Proponents of the new physics of the seventeenth century won out not because they explicitly refuted the arguments of the scholastics, but because the empirical success of their science rendered the scholastics' arguments irrelevant.

It may nevertheless help to eliminate some confusion and misunderstanding if we consider a few of the contemporary philosophical arguments against naturalism. Those arguments have mainly been provoked by W. V. O. Quine's spirited advocacy of "naturalized epistemology" of nearly two decades ago (Quine 1969). Many philosophers view the philosophy of science as largely a special application of epistemology. From this perspective, to naturalize the philosophy of science is to naturalize at least part of epistemology as well.

MUST A NATURALISTIC PHILOSOPHY OF SCIENCE DEFINE RATIONALITY?

Among the most ardent opponents of naturalized epistemology has been Quine's Harvard colleague, Hilary Putnam. One of Putnam's arguments, applied to a cognitive theory of science, would go as follows. A cognitive theory of science would require a definition of rationality of the form: A belief is rational if and only if it is acquired by employing some specified cognitive capacities. But any such formula is either obviously mistaken or vacuous, depending on how one restricts the range of beliefs to which the definition applies. If the definition is meant to cover *all* beliefs, then it is obviously mistaken because people do sometimes acquire irrational beliefs using the same cognitive capacities as everyone else. But restricting the definition to *rational* beliefs renders the definition vacuous. And so the program of constructing a naturalistic theory of science along cognitive lines goes nowhere (Putnam 1982, 5).

The obvious reply is that a naturalistic theory of science need not require any such definition. It is clear that Putnam is assuming a categorical conception of rationality. A naturalist in epistemology, however, is free to deny that such a conception can be given any coherent content. For such a naturalist there is only hypothetical rationality, which many naturalists, including me, would prefer to describe simply as "effective goal-directed action," thereby dropping the word 'rationality' altogether. Putnam's insistence that the naturalist in epistemology must provide a definition of categorical rationality in naturalistic terms is therefore just a disguised way of begging the question—of insisting that there must be a coherent concept of categorical rationality.

CAN INSTRUMENTAL RATIONALITY BE ENOUGH?

One way to challenge a naturalistic approach is to argue that a rationality of means is not enough. There must be a rationality of goals as well because, it is claimed, there is no such thing as rational action in pursuit of an irrational goal (Siegel 1985).

This sort of argument gains its plausibility mainly from the way philosophers use the vocabulary of "rationality." If one simply drops this vocabulary, the point vanishes. Obviously, there can be effective action in pursuit of any goal whatsoever—as illustrated by the proverbial case of the efficient Nazi. The claimed connection between instrumental and categorical rationality simply does not exist.

Moreover, investigating the actual goals of any group of scientists is an empirical matter, as is the investigation of the effectiveness of their means. That there is any way of evaluating the "rationality" of these goals, apart from considering other goals to which scientific activities might be an efficient means, is problematic. That such an evaluation *must* be possible is an unproven philosophical article of faith.

Nor does the restriction to instrumental rationality prevent the study of science from yielding normative claims about how science should be pursued. Indeed, it may be argued that the naturalistic study of science provides the only legitimate basis for sound science policy (Campbell 1985). What a naturalistic approach requires is that science policy be based on solid empirical findings about effective strategies for pursuing various scientific goals. If that requirement means the advice we may legitimately offer is of modest scope, that is all to the good.

Must the Naturalistic Study of Science Be Viciously Circular?

There is another long-standing objection to eliminating traditional epistemological questions in favor of questions about effective means to desired goals. To show that some methods are effective, one must be able to show that they can result in reaching the goal. And this requires being able to say what it is like to reach the goal. But the goal in science is usually taken to be something like "true" or "correct" theories. And the traditional epistemological problem has always been to *justify* the claim that one has in fact found a correct theory. Any naturalistic theory of science that appeals only to effective means to the goal of discovering correct theories must beg this question. Thus, a naturalistic philosophy of science can be supported only by a circular argument that assumes some means to the goal are in fact effective.

Before proceeding any further it is worth noting that, apart from some philosophers and sociologists of science, there is practically no knowledgeable group in the whole of Western (if not world) culture that seriously questions the obvious successes of science. Who would seriously question that Watson and Crick discovered the basic structure of DNA in the early 1950s? Or that geologists in the 1960s discovered that the crust of the earth consists of shifting plates? Or even that the structure of the poliovirus was determined in 1985? This is not to say that what most educated people believe is therefore true. But the disparity should give one pause. Why does the picture of science developed by philosophers and sociologists differ so greatly from that found among the educated public?

In the case of philosophers of science the answer must be sought in the history of what is still called "modern" philosophy—Descartes' seventeenth century program of universal doubt.

The Cartesian Circle

Descartes set out to provide a firm foundation for the developing science of his time by first doubting everything he could. In the course of rescuing himself from this self-created predicament, he argued for the existence of God on the basis of having a clear and distinct idea of a perfect being that necessarily exhibits the perfection of existence. Later Descartes attempted to justify his appeal to clear and distinct ideas by arguing that a perfect being would not allow humans to be deceived by a clear and distinct idea. But this justification, his critics objected, is circular. The objective is to give an argument that proceeds straight from some indubitable premises to the desired conclusion.

Since Descartes the standards for justification have been relaxed somewhat. Few would now require that the premises of the ultimate justifying argument be indubitable. And most would allow that the argument itself need not be strictly deductive. It is enough that the premises confer some appropriate degree of "probability" or "rational warrant" to the conclusion. Nor would many deny that most arguments in scientific practice use scientific premises. It is just that science is thought to require at least *some* arguments that justify their conclusions without the benefit of any premises themselves in need of such justification. How else, it is thought, could the whole scientific enterprise be rationally justified?[8]

The program of trying to justify science without appeal to any even minimally scientific premises has been going on without conspicuous success for 300 years. One begins to suspect the lack of success is due to the impossibility of the task. Perhaps there is simply no place totally outside science from which to justify science. At the very least one might conclude that the task is not going to be accomplished anytime soon by ordinary mortals. That would seem to be sufficient grounds for an ordinary mortal to try something else.

IS A NATURALISTIC PHILOSOPHY OF SCIENCE POSSIBLE?

It begins to look as if the answer to this question depends on how one defines the philosophy of science. If one adopts the standpoint of traditional epistemology, a naturalistic philosophy of science is impossible. Traditional epistemology insists on either solving the Cartesian riddle—refuting universal skepticism—or at least establishing a categorical conception of rationality. Nothing less will do. To pursue a naturalistic philosophy of science is to abandon these objectives.[9]

But one should not be deceived into thinking that the issue is merely a matter of definition. At stake is the autonomy of the philosophy of science and its assumed status as custodian and arbiter of scientific rationality. If the philosophy of science is naturalized, philosophers of science will find themselves on equal footing with psychologists, sociologists, and others for whom the study of science is itself a scientific enterprise. The most status a philosopher of science could claim is that of the theoretician of a developing cognitive science of science, on the model of theoretical, as opposed to experimental, physics. I am inclined to think that would be status enough.[10]

Evolutionary Naturalism

If one seeks a deeper foundation for the cognitive study of science, that can be provided. It lies not in epistemology, or in the philosophy of language, but in evolutionary theory.

Human perceptual and other cognitive capacities have evolved along with human bodies. We share many of these capacities with other primates and even lower mammals. Indeed, those parts of our brains responsible for our more advanced linguistic abilities are built upon and linked to those parts we share with other mammals. There can be no denying that these capacities are fairly well adapted to the environment in which they evolved. Without considerable adaptation, we would not be here. Nor are these capacities trivial. The perceptual and neural processing required just for a human to walk without falling or bumping into things is fantastically complex.

Empiricist philosophers emphasized the role of immediate perceptual experience in their analyses of knowledge because of the high degree of subjective certainty attached to such experience. Their problem was then to get beyond the immediate perceptions. From an evolutionary perspective the subjective certainty of perception is indeed causally connected with the source of the reliability of such judgments, which lies in our evolved capacities for interacting with our world. But the operation of these capacities is largely unrecorded in our conscious experience. Nor is there any obvious subjective trace of the evolutionary process that produced these capacities. Indeed, the biological nature and evolutionary origin of the connection between our subjective experience and our cognitive capacities is an interesting, and largely unsolved, scientific problem. The general *reliability* of the mechanisms that generate many of our perceptual and cognitive judgments, however, is, from an evolutionary perspective, not open to serious question.

Traditional rationalist philosophers, on the other hand, focused on our more general subjective intuitions, such as that space has three dimensions and that time exhibits a linear structure. These judgments seem to be built into the way we think. And indeed they are, for the aspects of the world relevant to our biological fitness have roughly that structure. But rationalists, like empiricists, could not see how to get beyond their subjective intuitions. They ended up in the ironic position of denying that the world (in which our capacities in fact evolved) is knowable by humans (that is, organisms possessing those evolved capacities). In proclaiming the subjective structure of our experience to be a necessary condition of all knowledge, they ruled out the possibility that we should ever be able to discover that the structure of the world as a whole in fact differs from that revealed by our intuitions—that, for example, space and time are really a four-dimensional space-time.[11]

At this point I am tempted to wax Wittgensteinian and suggest that traditional epistemology has been in the grip of a powerful picture, that of the "straight line" justification. It is this picture that makes the charge of "circularity" so powerful. Evolutionary theory provides an alternative picture. By looking back at evolutionary history, scientists themselves can better understand their own cognitive situation and investigate the development of their own cognitive capacities. What seem to the traditional epistemologist like vicious circles are, in this alternative picture, "positive feedback loops." Using our evolved cognitive capacities, we extend our knowledge of the world, including our knowledge of our own cognitive abilities. This latter knowledge helps us to extend our knowledge of the world still further. And so on. The existence of these loops is not a limitation that must be overcome by some special form of philosophical analysis. On the contrary, it is one of the things that makes modern science so powerful.

RATIONALITY IN BIOLOGICAL PERSPECTIVE

Thinkers struggling to understand the nature of their own knowledge in the seventeenth and eighteenth centuries may be forgiven for not having appreciated the evolutionary point of view. A century after Darwin a similar lack of appreciation is less forgivable.

The attempt to understand science in terms of a categorical conception of rationality fits squarely within the older biological tradition of attempting to understand species in terms of essential properties. The essential properties define an ideal type, and divergences from this ideal type are regarded as imperfections. Darwin taught us that there are no essential properties of organisms. There are only populations of individuals with variations in properties. Moreover, the variations are not to be lamented; they are necessary for evolution. They provide the material on which selective mechanisms operate.

The scientific world has long since given up essentialism in biology. It is part of a naturalistic approach to the study of science that we should at long last give up essentialism in epistemology as well. There is no point in seeking to characterize the ideally rational scientist. Rather, we should seek to explain the evolution of science in terms of the selective mechanisms operating on natural variations among real scientists.

CAN SOCIOBIOLOGY HELP?

The capacities evolution favors, of course, are just those that confer biological fitness, that is, the ability to survive and leave behind offspring. The ability to do modern science had nothing to do with the evolution of our perceptual and cognitive capacities. The general problem faced by a naturalistic philosophy of science, then, is to explain how creatures with our naturally evolved capacities have managed to fathom so much of the detailed structure of the world—"of atoms, stars, and nebulae, entropy and genes" (Gamow 1947, i).

Some sociobiologists have recently suggested that this task may be easier than one might at first think. They suggest that humans evolved special capacities, described by "epigenetic rules," for making typically scientific inferences. Hypothesizing a "common cause" on the basis of observed "coincidences" is one such suggested rule (Lumsden and Wilson 1981; Ruse 1986).

Although the existence of such epigenetic rules might strengthen the case for a cognitive theory of science, their status is currently controversial. Accordingly, I will make no appeals in this direction. But neither would I rule out the possibility that such capacities do exist.

EVOLUTIONARY EPISTEMOLOGIES

Up to now my appeal to evolutionary theory serves the following purposes. It provides an alternative foundation for the study of science. It explains why the traditional projects of epistemology, whether in their Cartesian, Humean, or Kantian form, were misguided. And it shows why we should not fear the charge of circularity.

A more ambitious use of evolutionary theory goes under the name of evolutionary epistemology. Evolutionary epistemologists hold that evolutionary theory itself provides a good model for the overall development of scientific knowledge. A strictly Darwinian account, for example, would try to show how scientific knowledge evolves through some mechanism of random variation and selective retention.[12]

Whether evolutionary models fit the development of scientific knowledge in interesting ways is an empirical matter. I would not myself approach the study of science with evolutionary models in hand, hoping to make them fit somehow or other. Nevertheless, the power of evolutionary models is considerable, and I shall not hesitate to employ them whenever they seem fruitful.

MIGHT EVOLUTIONARY NATURALISM IMPLY
A SOCIAL THEORY OF SCIENCE?

I have claimed that evolutionary naturalism supplies an empirical foundation for the cognitive study of science, which in turn is incompatible with a thoroughgoing social theory of science. But some social theorists would argue the reverse. Evolutionary naturalism, they would claim, requires a social theory of science.[13]

Imagine two research groups with strongly opposing theories. Surely the range of basic cognitive abilities in the two groups does not differ significantly. And they both have access to the same evidence. What explains the dispute must therefore be social in nature.

The general form of reply is this. The mere existence of evolved cognitive capacities does not determine the uses to which they are put. That depends on many things, including, of course, the social context. Thus, although one may need to invoke social factors to explain why scientists use their cognitive capacities in some ways rather than others, one must still appeal to their cognitive abilities to understand what they are doing when they employ those abilities in the specified ways.

The two competing research groups in the example may not differ in the basic cognitive abilities of their members, but they probably do in the kinds of models with which members of the two groups are comfortable. And they may differ in the range of experimental skills possessed by different members. These differences in *acquired cognitive resources* may well be an important part of the explanation of why the two groups disagree on which model is best.

In general, a cognitive theory of science need not deny that interests and other social factors may be important for explaining science. Indeed, it should provide a framework for understanding the role those factors do play. What it should deny is that social factors do *all* the important explanatory work.

What Might a Cognitive Theory of Science Be Like?

Developing a scientific theory of science is a reflexive enterprise in the sense that one is practicing a form of the very kind of activity under study. Thus, one necessarily begins with some commitments about one's subject embedded in one's own practice. This need not lead to paradox or irredeemable bias so long as one is able to change one's practice in the light of one's own findings. But because the mechanisms of self-correction may be fairly inefficient, it is important that one begin with a good first approximation.

One common conception is that science is a search for "laws" in the form of universal generalizations. Thus, many students of the scientific enterprise have looked for laws of science or of scientific development. Kuhn (1962), for example, proposed the now famous "stage theory" of scientific development: preparadigm science, normal science, crisis, revolution, new normal science, and so on. More recently, a group of philosophers of science has embarked on a project of testing models of scientific change, including Kuhn's, against data derived from the history of science. An example of the several hundred general claims considered is: "A set of guiding assumptions is never rejected unless an alternative set is available" (Laudan et al. 1986, 166).

The recent sociology of science provides similar examples of the search for laws on which to found a theory of science. In the final chapter of their already classic sociological study of the development of radio astronomy in Great Britain, Edge and Mulkay (1976, 382) set out to compare their study with five other studies. They proceeded to list fifteen "factors in scientific innovation and specialty development" that are exhibited in at least one of the six studies surveyed. Several of those factors appeared in all six. The factors cited included such things as "marginal innovation" (the introduction of novel ideas or techniques by people at the margin of the profession), "mobility" (into or out of related specialties), and "creation of new journal." Although they did not describe their remarks this way, those sociologists were looking for some laws of innovation and specialty development such as: The formation of a new specialty tends to be accompanied by the creation of new journals.

It is appropriate to test this conception of a theory of science against the history of science, and the history of biology provides a useful starting point. In the nineteenth century some biologists looked for biological laws among existing species and in the fossil record. Bergmann's law, for example, states that among species of warm-blooded vertebrates, members of subspecies living in cooler climates are larger than those living in warmer climates.[14]

If these "laws" are to be regarded as universal generalizations, they hover between falsity and vacuity. If stated with sufficient precision to be informative, they are always subject to some exceptions, such as burrowing mammals. If stated with sufficient generality or qualification to avoid exceptions, they cease to be usefully informative. At best such "laws" express only rough statistical generalizations reflecting the average selective forces of different environments on similar populations of organisms.

The laws of scientific development noted above are like the nineteenth century laws of evolutionary development. They provide at most some rough generalizations covering a very restricted class of cases. The problem is that the development of science is too variable at the level of the types of factors surveyed. Just as nature may solve an evolutionary problem in many different ways, so may scientific development occur in many different ways.

Even viewed as rough statistical generalizations, the purported laws of scientific development cannot be regarded as a *theory* of science. They cannot provide genuine explanations of science as a human activity. At best they describe some empirical phenomena for which any genuine theory of science should provide an adequate explanation.

THE MECHANISMS OF SCIENTIFIC DEVELOPMENT

A theory of science, like the modern theory of evolution, must be based on models that capture the "deeper structure" of scientific development. Here it is useful to consider the evolutionary analogy in somewhat greater detail.

Standard evolutionary models have three components: random variation of traits among members of a population; differential fitness, relative to the environment, of organisms with different traits; and inheritance of traits. Any population of organisms satisfying these conditions will evolve, provided only that the environment is not so hostile as to preclude survival. Genetics provides the mechanisms underlying the required variation and inheritance. But even the full genetical theory of natural selection has little to say about how specific populations will evolve in particular environments. It does, however, tell us much about the *process* of evolution.

By appealing to the mechanisms of representation and judgment, one could begin to develop an analogous model for science. Variations in representations exist within any population of scientists. The scientific judgments of individual scientists act differentially on different representations. Surviving representations are passed on to later generations through teaching and apprenticeship. In general terms the analogy is this: cognitive processes are to the development of science as genetic mechanisms are to the evolutionary development of populations.

I will not be concerned in this work to develop the evolutionary analogy in detail. My primary concern is with the more specific cognitive processes of representation and judgment. Only in the last two chapters, where I deal with specific historical developments in nuclear physics and in geology, will I again invoke an evolutionary model.

A Role for History?

Up to now I have not discussed the history of science. This may seem surprising because, of all the disciplines concerned with science as a cultural activity, the history of science is now the most active, and in many ways the most exciting. But historians offer us no explicit *theory* of science. To be sure, historians do have theories of science, often borrowed from elsewhere, implicit in their practice. But those theories are not the focus of their inquiry.

What, then, is the role of history in the enterprise of constructing a theory of science? For philosophers of science concerned to defend a categorical theory of scientific rationality, the role has always been difficult to define. History provides us only *descriptions* of how science has been pursued. From descriptions we cannot derive substantive *norms* governing how science should be pursued. History, in this framework, can at best provide illustrations of science having been pursued as some proposed norms would prescribe.

Any naturalistic theory of science opens the door to the possibility suggested by Kuhn in the very first chapter of *The Structure of Scientific Revolutions* (1962). The history of science, he said, can provide empirical evidence against which naturalistic theories of science, like any scientific theories, can be judged. But this suggestion contains pitfalls as well as promises. In particular, much depends on what kind of a theory of science one seeks.

If one seeks even only rough generalizations regarding scientific practice or development, one will not find them in the writings of historians. Of course, one will find instances from which one might seek to construct generalizations. But the use of those instances will always be troublesome because one has no idea of the sampling method by which some instances were selected for detailed historical reconstruction. Strong biases may exist that militate against the selection of instances that would weaken the desired generalization.

A cognitive theory of science is by nature much closer than standard philosophical accounts to the kinds of data provided by traditional historians of science. Historians have always been concerned with the ideas and choices of individual scientists. These historical findings are easily recast in the language of representation and judgment. And the worry that historians may not have selected representative cases is largely avoided. We know on biological grounds that the processes of representation and judgment are similar for most scientists. It is only the specific content of their representations and judgments that varies.

Finally, I would hope that historians of science will find here something that can be of benefit in the writing of history. The models of representation and judgment adapted from the cognitive sciences have more structure than the simple "ideas" and "choices" of folk psychology. And evolutionary models of scientific development recommend some changes in the way the history of science is conceived. In particular, the traditional "scientific biography" may not be the natural unit for the history of science—as social historians have argued, though for different reasons. To see the processes of variation and selection at work, one must look at many members of a research community, including those who are unsuccessful. Individual lives, as reconstructed by the individuals themselves, as well as by historians, exhibit far more purpose or design than in fact exists in the evolution of science as a whole.

Enough moralizing. The important point is that a cognitive approach to the study of science provides a basis for fruitful relationships between the history of science and the theory of science.[15]

Overview of This Book

In the next chapter I review some alternative theories of science. Readers familiar with the various theories are advised at least to skim these sections, for my review is not neutral but reflects my own point of view. Seeing what I say about alternative approaches will help readers understand the approach I develop in the later chapters.

Chapter 3, Models and Theories, explores the nature of the representations used in science and presents what is, in effect, a theory of theories. The view developed is that theories are heterogeneous entities consisting of families of models together with claims about the sorts of things to which the models apply. This view is no longer particularly novel. One new element is the explicit linking of scientific models with the "schemata" of the cognitive sciences. Another is the form of argument, which proceeds through the examination of professional-level textbooks in classical mechanics. I do not deny that families of models can be characterized axiomatically. Those characterizations, however, play little role in the scientist's understanding or use of theories.

Chapter 4, Constructive Realism, attempts to characterize a brand of scientific realism. The key to the view is that the important relationship between models and the world is not a semantic relationship, such as truth, but similarity between two nonlinguistic entities, an abstract model and a real system. Truth plays only the relatively trivial role of being a relationship between language, which may be used to characterize models, and the models so characterized. This relationship basically is thus simply one of definition.

Chapter 5, Realism in the Laboratory, presents a case for concluding that scientists are at least sometimes correctly described as constructive realists. It proceeds not by attempting to refute the arguments of empiricist philosophers or recent constructivist sociologists, but by offering a detailed account of actual laboratory research in a cyclotron facility devoted to nuclear physics. The focus is not on nuclear theory, but on the technology employed in experimentation. The only plausible interpretation of such research, I claim, is that protons and neutrons, which are paradigm cases of "theoretical" or "constructed" entities, are in fact research tools. They are not in any useful sense "hypothetical" but, like amplifiers and magnets, are simply part of the technology required for doing research in nuclear physics.

Chapter 6, Scientific Judgment, has two parts, one negative and one positive. The negative attacks the idea that scientists are to be understood as Bayesian agents—probabilistic information processors. Here the appeal is not to logical argument but to recent research on human judgment. The evidence, I claim, shows that humans are not probabilistic information processors. The positive part of the chapter develops an alternative account of scientific decision making based on the notion of *satisficing*. This model provides a decision-theoretic foundation for "interest" theories of scientific judgment. It also shows how scientists can make realistically interpreted decisions about how well a model fits the real world.

Chapter 7, Models and Experiments, attempts to show that scientists do sometimes employ satisficing strategies in designing experiments and in using experimental data to decide which of several models best fits the real world. Here I examine an ongoing episode in nuclear physics, using data from my own interviews with a number of scientists, both experimentalists and theoreticians, involved in this research. This case also illustrates how the cognitive processes of representation and judgment operate to produce the evolutionary development of a scientific field.

Chapter 8, Explaining the Revolution in Geology, illustrates how the model of science developed in the previous chapters can be used to illuminate historical episodes in science. Its main lesson is that data can sometimes be so clean, and models so well articulated, that scientists following a satisficing strategy have little freedom in which model to choose. Their other "interests," even if very strong, can be overwhelmed by their epistemic interest in choosing the "correct" model. Once again cognitive processes are shown to produce variation and selection leading to the evolution of a science.

The Epilogue, Reflexive Reflections, briefly examines the lessons of my theory of science for its own future development.

2

Theories of Science

Although the pictures of the philosophy and the sociology of science presented in the previous chapter are useful for highlighting a cognitive approach, they are not adequate for an informed comparison. A somewhat more detailed discussion of alternative views is thus required. But since it would be impossible to survey every account in a single chapter, I shall be quite selective both in the number of accounts treated and in the range of issues considered. I treat only the most actively discussed alternatives. And I focus primarily on issues related to the nature of scientific *representations* (beliefs, models, theories) and to scientific *judgment* (inductive logic, theory choice). These are the issues most relevant to a cognitive alternative.

With hindsight we can divide the twentieth century work in both the philosophy and the sociology of science into two periods separated by the unfolding influence, in the 1960s, of Kuhn's *Structure of Scientific Revolutions* (1962). Accordingly I begin with a look at two theories of science developed by philosophers and sociologists of science before the 1960s. After a brief presentation of Kuhn's own theory, I then examine several major post-Kuhnian theories of science.

Logical Empiricism

In 1960 logical empiricism was *the* Anglo-American philosophy of science. It had no serious rivals. Within the philosophy of science it still provides the primary foil for current investigations.[1]

The personal and social origins of logical empiricism are only now being investigated. The main strands of its *intellectual* heritage, however, are fairly clear. They are three. First was the mathematical and logical work of Hilbert, Peano, Frege, and Russell. This work in the foundations of mathematics provided both the model and the methods for logical empiricist studies of science. Second was the classical empiricism of Hume as transmitted through Mill to Russell and Mach. From empiricism came the presumption that experience, or "observation," provides the foundation for all scientific knowledge. The name "logical empiricism" well describes these first two intellectual sources of the movement. Third was science, which primarily meant the physics associated with Einstein—relativity theory and quantum mechanics.[2] The most enduring examples of logical empiricist analyses of actual science focus on these theories (Reichenbach 1928, 1958; Grunbaum 1963; Friedman 1983).

FOUNDATIONISM

Logical empiricism was explicitly "foundationist." Indeed, it was so in several different respects. First of all it was *conceptually* foundationist in the sense of the logicist program of *Principia Mathematica* (Whitehead and Russell 1910–13), which sought to "reduce" arithmetic to logic. Here one begins with logical notions like that of a set and membership in a set and then logically constructs entities with the formal properties of the integers. Having constructed integers, one uses them to construct rational numbers, and so on. The program assumes that the original logical notions are somehow more basic and conceptually clearer than the derived mathematical notions.[3]

Russell himself saw clearly how to apply this method to develop a foundationist version of traditional empiricist *epistemology*. In *Our Knowledge of the External World* (1914) he began with the existence of "sense data" and proceeded to analyze ordinary objects like tables and chairs as sets (or sets of sets) of sense data. Carnap's *Logical Construction of the World* (1928, 1967), which was directly inspired by Russell's work, was an attempt to develop Russell's suggestions in greater detail.[4]

Following Russell's program for the foundations of arithmetic, the logical empiricists conceived similar programs for the foundations of geometry, of physics, of biology, of psychology, and of sociology. The optimism and audacity of their program was astounding. They never seemed to doubt either that science *needed* philosophical foundations or that they possessed methods adequate to the task.

The logical empiricists had little interest in developing a *descriptive* account of how science actually works. Their aim was to provide science with "logical" and "epistemological" *foundations*. Their primary goal was to justify, or legitimate, science, not merely to explain how it works. But one cannot undertake the task of legitimization without having *some* picture of what it is one seeks to justify. The picture of science constructed by logical empiricists for the task of legitimization is the picture that dominated Anglo-American philosophy of science for a generation after World War II. One should not be surprised that a picture generated for the purpose of justification would turn out to be ill-suited for the purpose of description or explanation.

PSYCHOLOGISM

Frege, the codiscoverer of modern logic, taught that one should distinguish sharply between logic and psychology. Logic is not the science of thinking but, like mathematics, the study of purely logical relationships among objective, though abstract, entities. The supposed confusion of logic and psychology was labeled "psychologism." To commit psychologism was, for logical empiricists, a cardinal sin. The same would hold for "sociologism."

The application of Frege's teachings to science is immediate. The philosophical study of science is not the study of how scientists actually think or act. That is left to psychologists and sociologists. It is the study of logical relationships among scientific propositions. As Carnap (1937, xiii) put it:

> Philosophy is to be replaced by the logic of science—that is to say, by the logical analysis of the concepts and sentences of the sciences, for the logic of science is nothing other than the logical syntax of the language of science.

Since logic is a purely a priori discipline, it follows that the philosophical study of scientific thinking is an a priori endeavor. It is the study of how, a priori, an ideally logical scientist *should* think.

For logical empiricism, then, the gap between the psychology or sociology of science and the philosophy of science is like the gap between "is" and "ought." It is logically unbridgeable.

Reichenbach's much discussed contrast between the "context of discovery" and the "context of justification" is not an independent distinction; it follows directly from the more fundamental distinction between descriptive psychology and normative logic. By "discovery" the logical empiricists meant not the discovery of new phenomena but of new *hypotheses*. The context of discovery is the context of thinking up hypotheses. Here there is no logic, only psychology. Logic applies only to the question whether the resulting hypothesis is justified. A forceful statement of this doctrine appeared in Popper's *The Logic of Scientific Discovery*. The opening words of a section therein entitled "Elimination of Psychologism" read (Popper 1959a, 31):

The work of the scientist consists in putting forward and testing theories. The initial stage, the act of conceiving or inventing a theory, seems to me neither to call for logical analysis nor to be susceptible of it. The question how it happens that a new idea occurs to a man—whether it is a musical theme, a dramatic conflict, or a scientific theory—may be of great interest to empirical psychology; but it is irrelevant to the logical analysis of scientific knowledge.

THEORIES AS FORMAL SYSTEMS

The logical empiricists devoted much effort to developing an account of what a scientific theory is. Taking the foundations of mathematics as the model, they held that a theory is a formal, logical system. What distinguishes scientific theories from pure logic or mathematics is that the nonlogical terms are given an *empirical interpretation*. The symbol R in a formal axiomatic system, for example, might be interpreted as standing for the path of a light ray. But logical empiricists were not concerned with the actual process of interpretation. Rather, interpretation was represented by special "correspondence rules," or "meaning postulates," such as "R stands for the path of a light ray." These were typically understood as being among the statements that make up the theory of physical geometry.

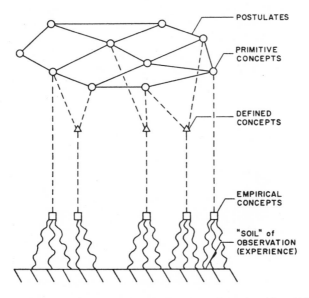

Figure 2.1. A logical empiricist picture of a scientific theory. Reproduced from Feigl (1970).

The empiricist commitments of logical empiricism gradually infected their view of scientific theories. In postwar writings, the vocabulary of a theory was divided into two parts, a "theoretical" part and an "observational" part. The standard doctrine was that the *meaning* of theoretical terms is totally a function of the meaning of the observational terms together with the purely formal relations specified by the axioms of the theory. The result is *instrumentalism* with regard to the theoretical part of theories. Theoretical terms do not refer to real entities; they are mere instruments for organizing claims about the things referred to by observational terms. The entities that one might uncritically suppose are referred to by theoretical terms are at most useful fictions. This view of scientific theories was typically pictured as in figure 2.1 (Feigl 1970).[6]

Not all logical empiricists were instrumentalists. Indeed, Reichenbach claimed to be a "realist." But it is difficult to find anyone associated with logical empiricism who did not take it for granted that scientific theories should be thought of as formal, axiomatic systems with some sort of empirical interpretation. This was the starting point for all further analyses of science.

INDUCTION AND PROBABILITY

By the mid 1930s the logical empiricists had given up the Russellian dream of logically constructing the body of physical theory from a base in observation and took up the problem of justifying inferences from observation to theory. In doing so, they exposed the Achilles heel of empiricism, Hume's problem of induction. In contrast to the near unanimity concerning the nature of scientific theories, their responses here differed in fundamental ways.

The program that best represented the ideals of logical empiricism was Carnap's (1950) attempt to construct an inductive logic that would be a strict generalization of Russell's deductive logic. Indeed, a similar attempt had been made by John Maynard Keynes (1921) just a few years after publication of *Principia Mathematica*.

The basic idea is simple. The "law statement," L, "All swans are white" deductively implies the statement "If Sam is a swan, Sam is white." What is desired is a logical system in which observational statements of evidence like, e, "Sam is a white swan" "partially imply" the law. Carnap, like Keynes before him, thought partial implication should be represented by a probability function, so that inductive support has a logical form represented by the probability statement $p(L/e) = r$ (read: The probability of L, given evidence e, has value r).

Never mind that Carnap developed his inductive logic only for languages far too simple to express even the most elementary claims of modern science. And never mind the many technical difficulties in making the system work at all. What is important is the vision of rational scientific inference as totally objective in just the way deductive logic was perceived to be. This vision kept philosophers working on variants of Carnaps's inductive logic well into the 1970s.[7]

Popper is famous for his long-standing polemics against any form of inductive logic. But his commitment to the ideal that scientific reasoning should be a matter of logic has been, if anything, stronger than Carnap's—and much less sophisticated. Scientific theories, he has always claimed, have the logical form of universal generalizations: "All As are B." The falsity of any such claim follows *deductively* from an evidence statement of the form "This A is not B." All that science requires, claimed Popper, is that observational evidence can thus logically falsify a scientific theory. The most that can be said in favor of so austere a view is that, by refusing to make any inductive inferences at all, it does avoid Hume's problem.[8]

The other giant of logical empiricism, Hans Reichenbach, took a different line altogether. Induction, he argued, is not a matter of logic at all but of *practical action*. His view was developed in detail only for claims about the limiting relative frequency of attributes in an infinite sequence—which have roughly the structure of probabilities. If a fraction, f, of observed swans have been white, we should, as a matter of practical action, "posit" that the limiting relative frequency of swans in the infinite limit will be f. The justification for this practice is pragmatic. It is that persistence in following this policy is logically guaranteed eventually to lead to a true posit so long as a true posit exists, that is, so long as the sequence in question does indeed have a limit.[9]

The idea that the real goal of science is to make claims about limits of infinite sequences seems bizarre. But Reichenbach is not the only philosopher to have advocated the view that truth in science can be guaranteed only in the infinite long run of inquiry. And again, one of the chief philosophical merits of such a view, and one that deeply concerned Reichenbach, is that it provides a modern, scientific answer to Hume. The justification of the inductive policy relies solely on a priori truths about the mathematical properties of infinite sequences—properties discovered only in the nineteenth century.[10]

A common thread in these otherwise very divergent approaches to scientific inference is that "rational" inference is to be understood and justified solely in terms of logical or mathematical categories. This is the belief, as Stephen Toulmin (1972) was later to describe it, that rationality reduces to logicality. The rejection of this belief is a hallmark of later philosophies of science.

SCIENTIFIC EXPLANATION

For those who first became familiar with the philosophy of science in the United States after World War II, the work most likely encountered was Hempel's theory of scientific explanation. Partly this is because Hempel applied his theory to functional explanations in biology and the social sciences, and to explanation in history. All told, Hempel's "Studies in the Logic of Explanation" (Hempel and Oppenheim 1948) has been one of the most discussed papers in the philosophy of science.[11]

Hempel's approach to the topic followed the logical empiricist guidelines noted above. His concern was not directly with how things are explained but with the "logic" of explanation. That logic turns out to be simple deduction. A scientific explanation, according to Hempel, is a deductive *argument* in which a statement describing the event to be explained (This raven is black) is deduced from a true general law (All ravens are black) together with appropriate initial conditions (This is a raven). Many years later he attempted to generalize his account to include "inductive statistical" explanations as well (Hempel 1962).

Since *prediction* in science was also analyzed as consisting of an argument from laws and initial conditions to a description of the predicted event, it seemed an immediate consequence of Hempel's analysis that there exists a logical symmetry between explanation and prediction. The only difference is the "pragmatic" one of whether one is applying the argument before or after the event in question has taken place.

THE LEGACY OF LOGICAL EMPIRICISM

There are today only a few philosophers of science who would defend any major logical empiricist doctrines in anything like their original form. Yet the legacy lives on in assumptions held even by those most opposed to the logical empiricist account of science. One is the widespread presumption that scientific theories are essentially linguistic entities—maybe not formal axiomatic systems, but nevertheless something like sets of statements. A second is the presumption that the job of the philosophy of science is at least to "explicate," even if no longer to justify, the rules of rational belief and action that are supposed somehow to "govern" scientific activity. Of course, some philosophers have challenged even these presumptions. A cognitive approach to the study of science, as I understand it, rejects both. And it does so in a theoretically systematic way.

The Social Structure of Science

The empirical psychology of science that logical empiricists assumed would complement the logic of science did not materialize. In 1960 only a few scattered works existed that fit clearly under the banner "psychology of science." Ironically, part of the reason for this failure was that the behaviorist psychology championed by logical empiricists did not have the resources to deal with so complex a cognitive activity as science.

A field which did emerge to complement the logic of science was the sociology of science. Its rise, at least in the United States, was primarily due to one man, Robert K. Merton. Merton's scholarly career began with his classic work on science, technology, and society in seventeenth century England (1938). Later his interest in science was more as a strategic research site for broader interests in sociology. There is little evidence that he ever consciously thought of his work as complementary to the program of logical empiricism. But his program was complementary nonetheless.[12]

In the opening section of a very influential early (1942) paper, Merton (1973, 268) wrote:[13]

> We are here concerned in a preliminary fashion with the cultural structure of science, that is, with one limited aspect of science as an institution. Thus, we shall consider, not the methods of science, but the mores with which they are hedged about. To be sure, methodological canons are often both technical expedients and moral compulsives, but it is solely the latter which is our concern here. This is an essay in the sociology of science, not an excursion in methodology. Similarly, we shall not deal with the substantive findings of sciences (hypotheses, uniformities, laws), except as these are pertinent to standardized social sentiments toward science. This is not an adventure in polymathy.

The complementary position to the logical empiricist strictures against sociologism could hardly have been better expressed. Though not explicitly, and probably not even consciously, Merton left the analysis of "methodology" and the "substantive findings of science" to the logical empiricists. He was providing an analysis of the "cultural structure" of science, which as it turned out complemented their analysis of its "logical structure."

THE NORMS OF SCIENCE

What, then, are the primary characteristics of Merton's theory of science? It is, first of all, an application of *functional analysis*. In Mertonian terms this means investigating the specific functions of various social norms that constitute a social structure. His 1942 summary of the nature of social norms in science went as follows (Merton 1973, 268–69):

> The ethos of science is that affectively toned complex of values and norms which is held to be binding on the man of science. The norms are expressed in the form of prescriptions, proscriptions, preferences, and permissions. They are legitimatized in terms of institutional values. These imperatives, transmitted by precept and example and reenforced by sanctions are in varying degrees internalized by the scientist, thus fashioning his scientific conscience.

The social norms of science, then, are presented as functioning like the social norms operative in any cultural institution.

In this early essay Merton focused on four specific "sets of institutional imperatives" he thought to be characteristic of science as a cultural institution. These were universalism ("The acceptance or rejection of claims entering the lists of science is not to depend on the personal or social attributes of their protagonist"); communism ("The substantive findings of science are a product of social collaboration and are assigned to the community"); disinterestedness ("The demand for disinterestedness has a firm basis in the public and testable character of science"); and organized skepticism ("The temporary suspension of judgment and the detached scrutiny of beliefs in terms of empirical and logical criteria"). These capsule summaries of the four norms hardly do justice to Merton's work, but they give the flavor of his enterprise.

The existence of specific norms, and their maintenance in a social institution, is explained by the *function* that norms play in furthering the goals (values) of the institution. Turning once again to the 1942 essay (Merton 1973, 270):

> The institutional goal of science is the extension of certified knowledge. The technical methods employed toward this end provide the relevant definition of knowledge: empirically confirmed and logically consistent statements of regularities (which are, in effect, predictions). The institutional imperatives (mores) derive from the goal and the methods. The entire structure of technical and moral norms implements the final objective.

The existence and maintenance of the four norms, then, would ultimately be explained by their function in furthering the acquisition of "certified knowledge."

In one respect the approaches of logical empiricism and of functionalist sociology are more than complementary. Both schools assume that human action is rule governed. In the one case, cognitive activities are said to be governed by rules of rationality; in the other, social activities are said to be governed by institutionalized rules of social action. This assumption is now being challenged in both spheres. Even if the existence of such rules is acknowledged, their role in determining individual actions, or even in just legitimating these actions, is being disputed. Consequently the appeal to norms, whether of rationality or of action, is now thought by some to provide little explanation of scientific beliefs and actions.[14]

THE REWARD STRUCTURE OF SCIENCE

According to one sympathetic commentator, Norman Storer, the Mertonian "paradigm" for a sociology of science became established only with the publication of Merton's 1957 essay "Priorities in Scientific Discovery" (Merton 1973, 286–324). "The basic idea of interaction between the normative structure and the reward structure of science," Storer wrote, as editor of Merton (1973, 283), "provides a solid foundation for the understanding of science as a social institution."

In that 1957 paper Merton investigated a problem he first noted in his doctoral dissertation. Why are "priority disputes" in science so common and so vitriolic? And why are they often pursued by scientists not directly involved? His answer, in a nutshell, is this.

The creation of new, certified knowledge is the goal of the institution of science. To encourage scientists to pursue this goal, it is functional for research communities to create a reward structure that generally rewards original results. One of the main forms of reward in science is recognition of scientific achievement. A norm of scientific research is, therefore, that those who produce original work should be rewarded with appropriate recognition. To make the reward structure work, it is necessary to determine who did the original work, that is, to give credit where it is due. Given the possibility of nearly simultaneous discoveries, indeed the probability of such discoveries assuming the norm of communism, priority disputes are to be expected. Their intensity, and the fact that they are pursued by other parties, indicates the importance of the norms and values behind the reward system. Here, then, is an example of the kind of explanation that is to be expected from a functionalist study of the social structure of science.[15]

COGNITION AND SOCIAL STRUCTURE

In separating the "substance" and "methodology" of science from its social structure, functionalist sociology cut itself off from any explanation of the distinctive cognitive products of science. Worse yet, any functional explanation of the social norms of science must *assume* some understanding of what constitutes "certified knowledge" and of the "methodology" that yields such knowledge. If it does not, the functional connection between the social structure and the goals of the institution cannot be fully understood. Yet the picture of the internal workings of science assumed in the functionalist literature seems to be even less adequate than that provided by logical empiricism. Little wonder that recent sociologists of science concerned with the *content* of science have turned their backs on the functionalist approach.

Paradigms and Revolutions

With a quarter century of hindsight it is now clear that Kuhn's *Structure of Scientific Revolutions* (1962) is, by any measure, the most influential book on the nature of science yet to be published in the twentieth century. It was a major contributor to the decline of logical empiricism beginning in the 1960s and, a decade later, to the decline of functionalist sociology of science. Of course, already in 1962 other works even more directly critical of logical empiricism had been published, including some appealing to the history of science. But Kuhn was the only theorist at the time to provide an alternative overall framework in which to investigate the nature of science.[16]

There is no point here even to attempt a summary of the vast secondary literature on Kuhn's book. Instead, I shall offer my own interpretation of his theory, one that emphasizes its naturalistic and cognitive aspects. Kuhn himself has remarked that his book "can be too nearly all things to all people" (1974, 459). To me it is an early attempt to develop a cognitive theory of science.

KUHN'S NATURALISM

N. R. Hanson is a prime example of a philosophical critic of logical empiricism who predated Kuhn. He asked (1958, 1961), "Is there a logic of *discovery?*" He answered in the affirmative, using historical cases. Kuhn's position was far more radical. His implied question was, "Is there a logic of *justification?*" And his answer was, "No!"

That this really was Kuhn's position comes out clearly in the very first chapter of *The Structure of Scientific Revolutions*. Describing Section XII, "The Resolution of Revolutions," he wrote (1962, 8):

> Section XII describes the revolutionary competition between the proponents of the old normal-science tradition and the adherents of the new one. It thus considers the process that should somehow, in a theory of scientific inquiry, replace the confirmation or

falsification procedures made familiar by our usual image of science. Competition between segments of the scientific community is the only historical process that ever actually results in the rejection of one previously accepted theory or in the adoption of another.

In short, there is no "logic" of justification. There is only the historical process of competition among segments of a scientific community.

Nor was Kuhn ignorant of the conflict between his views and those of contemporary philosophers of science. The paragraph immediately following the above passage reads (1962, 8–9):

> Undoubtedly, some readers will already have wondered whether historical study can possibly effect the sort of conceptual transformation aimed at here. An entire arsenal of dichotomies is available to suggest that it cannot properly do so. History, we too often say, is a purely descriptive discipline. The theses suggested above are, however, often interpretive and sometimes normative. Again, many of my generalizations are about the sociology or social psychology of scientists; yet at least a few of my conclusions belong traditionally to logic or epistemology. In the preceding paragraph I may even seem to have violated the very influential contemporary distinction between "the context of discovery" and "the context of justification." Can anything more than profound confusion be indicated by this admixture of diverse fields and concerns?

And referring back to a variety of standard philosophical distinctions, including that between "discovery" and "justification," he replied (1962, 9):

> Rather than being elementary logical or methodological distinctions, which would thus be prior to the analysis of scientific knowledge, they now seem integral parts of a traditional set of substantive answers to the very questions upon which they have been deployed. That circularity does not at all invalidate them. But it does make them parts of a theory and, by doing so, subjects them to the same scrutiny regularly applied to theories in other fields. If they are to have more than pure abstraction as their content, then that content must be discovered by observing them in application to the data they are meant to elucidate. How could the history of science fail to be a source of phenomena to which theories about knowledge may legitimately be asked to apply?

Although he did not use exactly these words, Kuhn was here advocating a *naturalized* philosophy of science. And although he seems deliberately to have avoided the phrase, his was an excursion into what writers before and since have called "the science of science." Little could have been further from the intentions of most philosophical critics of logical empiricism. And nothing short of metaphysics could have been more directly opposed to the program of logical empiricism.

KUHN'S THEORY OF SCIENCE

Despite the word "structure" in its title, Kuhn's book is not about the structure of scientific revolutions in anything like the way Carnap's work is about the structure of scientific theories or Merton's is about the social structure of science. Kuhn's is a theory of scientific *development* in the way Marx's was a theory of economic development or Piaget's a theory of cognitive development. And as did Marx and Piaget, Kuhn offers us a series of stages. Indeed, the stages provide the chapter-by-chapter organization for Kuhn's whole book.

Now, students of both Marx and Piaget have long since concluded that what is important about their theories is not the progression of stages. There are just too many examples where the actual progression does not follow the prescribed pattern. What is important is the different mechanisms and types of activity operating in different periods. The stages merely provide a loose way of organizing discussion of these topics.

Even downplaying Kuhn's stages, to say that he has offered us an account of the development of *science* is still too broad. Science is too large a unit of analysis. What, then, is the working unit whose development Kuhn delineates? The answer is clearer from Kuhn's examples than from the text. The unit of analysis is a scientific area, field, or specialty. Among his many examples are mechanics, optics, electricity, chemistry, geology, and genetics. It is such fields that undergo development from preparadigm science, to normal science, to revolutionary science, to normal science, and so on.

The fundamental notion in Kuhn's theory is that of a "paradigm" as an exemplary solution to a specific problem—a solution that can be used as a "model" to fashion solutions to other problems in the same field. Newton's solution to the problem of orbital motion and Fresnel's solution to the problem of diffraction are among his favorite examples. For Kuhn, such solutions are not to be viewed as applications of general theory, much less as tests of general theory. They are, on the contrary, the basic materials of which a normal science research tradition, which also includes such things as instrumentation, is constructed. That exemplars are logically prior to general theory is what Kuhn meant by "the priority of paradigms" (1962, chap. 5).

Kuhn himself now agrees with his many critics that he originally overused the idea of a paradigm, allowing it to play too many different roles. The other major use of the term was to designate a quite inclusive entity including instrumental techniques, general theory, and even a metaphysical world view. It is in this more global sense of 'paradigm' that Kuhn speaks of "the Newtonian paradigm" in mechanics, or even all of physics, throughout the eighteenth century. There can be no doubt that paradigms in this sense do exist, and it is convenient to have a term to describe them. But I think the term in this sense plays no real role in Kuhn's theory of science that is not covered by his account of a normal science research tradition. That account was one of Kuhn's most important contributions to the theory of science.[17]

A striking feature of normal science is that it is *not* an enterprise of testing general theory. Rather, according to Kuhn, it is a process of extending the techniques associated with the original exemplars to new problems. He calls it "puzzle solving." He also calls it "articulating" the paradigm. In any case, scientists are not asking whether the tradition is right or wrong. The question is whether they are clever enough to solve the new puzzles.

The main result of introducing the global sense of 'paradigm' was to create confusion both in Kuhn's own mind and in those of his critics. As one example it led Kuhn to identify the transition from pre-science to science as a transition from preparadigm research to paradigm-based research. Yet by his own account the prescientific period would be better described as one in which several normal science traditions operate simultaneously in a single domain. Each has its own exemplars used to guide further research. Yet for some reason—perhaps because the problems suggested by the various exemplars do not sufficiently overlap, or perhaps because no one tradition is very well developed—competition remains indecisive. Kuhn may be right that progress in solving problems improves when all the effort is guided by a single set of exemplars. But how and why the multiple traditions give way to a single tradition has not been adequately explained. Perhaps this question has not seemed very acute because it is a less dramatic version of the question how and why one normal science tradition gives way to another. That is the problem that excited philosophers of science.

The global sense of 'paradigm' misled Kuhn's philosophical critics because they generally took it to mean *theory,* in some suitably fuzzy sense. They then reduced Kuhn's questions about the nature of scientific revolutions to questions about the choice of one theory over another—a topic philosophers have long debated. But the question in Kuhn's account is: What leads a scientific community to abandon one set of exemplars (and thus one research tradition) for another? Since research traditions somehow include theories, the philosophers' question is included. Kuhn, however, says very little about theories as such. His theory of science does not include an explicitly developed theory of theories. Kuhn's question is importantly different from the philosophers' substitute.[18]

How then do revolutionary changes occur? Here the basic notions are "anomaly" and "crisis." An *anomaly* is a puzzle that no one has succeeded in solving, even though many may have tried and it appears that the problem ought to be solvable using techniques modeled on established exemplars. One of Kuhn's major points is that anomalies exist in all fields at all times. If paradigms were theories, and applications deductions from theory, all theories would have to be regarded as falsified—and presumably rejected. But, contrary to Popper's philosophy, this is obviously not so. Anomalies alone, therefore, do not produce scientific revolutions.

Another "necessary precondition" for revolution is the existence of a sense of *crisis*. This happens when, although solving the anomalous puzzle is regarded as necessary for further progress in the field, many practitioners become convinced that no articulation of standard exemplars will do the trick. Radically new solutions, which is to say, new exemplars, are needed. The search for new exemplars ushers in a period of "revolutionary science." Just what produces a sense of crisis is not clear from Kuhn's account.

One might think that what it takes to end the period of revolutionary science, and thereby establish a new normal science tradition, is good solutions to the problems that provoked the crisis. For Kuhn, however, the process cannot be so simple because research traditions are said to be necessarily "incommensurable." What Kuhn meant by 'incommensurability' is one of the major debates in Kuhnian scholarship.

Indeed, there are several kinds of incommensurability. One is incommensurability of *meaning*. Kuhn said that in the revolutionary transition from one normal science tradition to another, words change their meaning. 'Mass' in Einsteinian physics, he said, does not mean the same thing as 'mass' in Newtonian physics. And when he said the practitioners of different normal science traditions tend to talk past one another, he seemed to be saying that one does not fully understand what the others are saying.

It was incommensurability of linguistic meaning that most philosophers took as fundamental to Kuhn's account. Meaning, after all, has long been a focus of philosophical interest. And philosophers such as Kripke (1972) and Putnam (1975) invented a new theory of reference to meet the challenge. But incommensurability of meaning, I now think, was not the major problem. The key to an adequate understanding of science is not to be found in the philosophy of language.

The real problem in comparing different normal science traditions is incommensurability of *standards*. In the process of creating a normal science research tradition, a scientific community develops "norms," "rules," or "standards." These help determine, among other things, what counts as a significant problem and what counts as a solution. But Kuhn's doctrine of the priority of paradigms (that is, exemplars) holds that the standards are determined by the exemplars, not the exemplars by the standards. It follows that a new solution to a problem, one appealing to a new exemplar and thus violating the standards of the older research tradition, could not be recognized as a solution by that tradition. A solution to the problem of atomic spectra that invokes quantized values for the exchange of energy, for example, could not be a solution within a classical tradition that presumed continuity. In recognizing a new solution as indeed a solution, therefore, scientists are creating a new tradition, and thus new standards. But their prior decision to accept the new solution is then not governed by any standards at all. It is a truly revolutionary act in the classical, political sense.[19]

The problem would disappear, of course, if there were standards of choice independent of any particular normal science research tradition. But this Kuhn (1962, 93) denies:

> As in political revolutions, so in paradigm choice—there is no standard higher than the assent of the relevant community.

One can only study the various processes by which scientists finally reach agreement on a new set of exemplars. These include techniques of persuasion and the gradual dying off of adherents to the older tradition.

RELATIVISM AND COGNITIVISM

It has been a common charge among philosophers of science that by making all scientific standards internal to a normal science research tradition, Kuhn rendered his account of science fatally relativistic. To avoid such relativism, it is typically assumed that there must be content-neutral standards that could be applied to any normal science tradition.

There is, however, a way to avoid relativism without appeal to "standards" at all. This is to focus on *cognitive processes,* such as those involved in representation and judgment, which are shared by all scientists. To some extent, Kuhn recognized this option. He appealed to the notion of a "gestalt switch" to explain how scientists change from one tradition to another—how they come to see a solution as an exemplar. Obviously the capacity to make this change does not depend on prior adherence to any normal science tradition. It is part of the natural, cognitive endowment of all scientists.

Unfortunately the capacity to experience gestalt switches was just the wrong kind of cognitive capacity on which to construct a bulwark against relativism. The primary characteristic of such switches is that they involve a change in perception with *no* corresponding change in the world. But such cognitive capacities are hardly exhaustive of all human capacities for interacting with the world on a cognitive level. By investigating some other capacities, one might see why a relativism of research traditions need not follow from an account of science in which exemplars are primary and overarching standards of rational choice are nonexistent.[20]

Programs and Traditions

Among Kuhn's many philosophical critics Imre Lakatos and Larry Laudan have attracted the most attention. Each has developed an alternative account of science that agrees with Kuhn in being historically developmental but disagrees by being an account of *rational progress* rather than of mere natural development.[21]

THE METHODOLOGY OF SCIENTIFIC RESEARCH PROGRAMS

Lakatos's methodology of scientific research programs is an audacious mixture of ingredients borrowed from Popper, Kuhn, and Quine, perhaps, as Hacking (1979) has argued, with a pinch of Hungarian-style Hegel and Marx. Lakatos himself presented his theory as an improvement on Popper's falsificationist methodology, which Popper had presented as an improvement over both "inductivism" and "conventionalism" (Lakatos 1970, 1971).

Lakatos began by agreeing with Popper that the fundamental problem for a theory of methodology is to provide a *demarcation* between science and nonscience. Popper had proposed falsifiability as the criterion of demarcation. But Lakatos took to heart Kuhn's point that, as he colorfully reformulated it, all theories live in an "ocean of anomalies" (1970, 135). By interpreting an anomaly as a false consequence of a theory, Popper, according to Lakatos, would have had to conclude that all theories are falsified and should therefore be rejected. But this, as Kuhn repeatedly illustrated, is manifestly not what happens even in the primary examples of great science. Popper's criterion of demarcation, Lakatos concluded, is too strong.

In response Lakatos embraced the Duhem-Quine thesis that falsification of any theory can always be avoided by invoking suitable "auxiliary hypotheses." But this point threatened to land him back in "conventionalism," which he, like Popper, rejected. To avoid conventionalism

Lakatos introduced a variant on Kuhn's idea of a normal science tradition—interpreted, now, as a succession of theories together with methodological rules (1970, 132):

> It is a succession of theories and not one given theory which is
> appraised as scientific or pseudo-scientific.

A succession of theories cum methodological rules is a "research program."
   More exactly a Lakatosian research program has three components. First
is a theoretical component consisting of a "hard core" of laws plus a "pro-
tective belt" of "auxiliary hypotheses." Like the logical empiricists and
Popper, Lakatos thinks of individual theories as sets of statements. Kuhn's
"concrete paradigms," or exemplars, play no role. The second component
is a "negative heuristic," a methodological rule forbidding the use of
counter-instances to refute statements in the hard core. Refutation is to be
avoided by suitable additions to or subtractions from the set of auxiliary
hypotheses. The third component is a "positive heuristic," which lays out
directions for future development and provides suggestions as to what sorts
of new auxiliary hypotheses might be fruitful.
   Like Kuhn, Lakatos found support for his view in the history of Newtonian mechanics. Newton's three laws together with the law of universal
gravitation form the hard core. When anomalies, such as the motion of the
moon, threaten, the core is not rejected. Instead, resources implicit in the
hard core, such as forces and angular momentum, can be used to fashion
additional auxiliary hypotheses until observation comes to agree with the
hard core.
   Lakatos appraised programs, not theories, as being either "progressive"
or "degenerative." To be progressive, a program must use its positive
heuristic successfully to predict new, or novel, phenomena. Here Lakatos
followed Popper in assuming the scientific value of increased truth content.
A program that can succeed in accounting for phenomena by inventing
auxiliary hypotheses only after the fact is said to be degenerative. Here
Lakatos agreed with Popper's distaste for ad hoc maneuvers. But not all
predictions need be successful. In fact, not even a majority. In this respect
Lakatos turned Popper's falsificationist methodology on its head. For
Popper, confirmations count for little; refutations are what matter. For
Lakatos, refutations count for little; confirmations are what matter.
   In Lakatos's theory, competition in science is between (or among) re-
search programs. A revolution is merely a case of a newer program super-
seding an older one. For Kuhn there is no independent question whether
the change constitutes progress. There is nothing more to progress than

SIENA COLLEGE LIBRARY

gaining the allegiance of the majority of scientists in the field. That Lakatos called "irrationalism." His is a theory of "rational progress" (1970, 93), which occurs only if the successful program is in fact more progressive than previous ones by the standards of Lakatos's methodology.

Lakatos therefore sought to reinstate the distinction, fundamental to logical empiricism, between the "logic" of science and its psychology or sociology. He made the distinction coextensive with that between "internal" and "external" history. If a program that suceeds is more progressive than its rivals by Lakatos's standards, only the internal history of that episode, told in terms of the categories provided by the methodology of research programs, is required to explain the episode as part of the history of science. Inquiry into the social or psychological facts of the case would be superfluous. Only if the historically successful program is *less* progressive than its rivals by Lakatos's standards is further explanation called for. We must then turn to psychology and sociology to explain why the scientific community did not follow the path of "rational progress."

Kuhn, who recognized no standards beyond those embodied in a particular research tradition, might well question the basis for Lakatos's claim that his account of rational progress is the right account. To his credit Lakatos recognized this type of objection and attempted to meet it. He coined the term 'metamethodology' to refer to the "quasi-empirical" method by which methodological research programs are themselves judged as progressive or not. The method, implicit in his criticism of Popper, is simply to see which methodological account "can reconstruct more of actual great science as rational" (1971, 117). I will consider the prospects for "metamethodology" below.

PROBLEMS AND PROGRESS

Laudan's *Progress and Its Problems* (1977) was an attempt to develop "a potentially more adequate model of scientific rationality" (1977, 4), that is, a better model of science than those presented by the logical empiricists, Kuhn, or Lakatos. Like Kuhn and Lakatos, Laudan focused on the *research tradition* as "the primary unit of rational analysis" (1977, 5). Like Kuhn, but unlike Lakatos, he took *problem solving,* not the generation of novel content, to be the primary goal of a research tradition. Unlike both Kuhn and Lakatos, he recognized *two* fundamentally different categories of problems.

The standard kinds of problems, which Laudan called "empirical," are generated by empirical findings that have no explanation in terms of current theory or, more seriously, seem contrary to current theory. *Anomalies,* for Laudan, arise in a given research tradition only when some theory in a *rival* tradition solves the problem. A problem is "solved" by theory T "if T functions (significantly) in any schema of inference whose conclusion is a statement of the problem" (1977, 25).

Laudan called the second category of problems "conceptual" and catalogued several varieties. Conceptual problems are, above all, problems "exhibited by some theory or other" (1977, 48). Conceptual problems may be internal, as when a theory is suspected of inconsistency. Or they may be external, as, for instance, if the theory in question conflicts with another theory or with more general "metaphysical" views. The difficulty of reconciling Newton's gravitational "action at a distance" with the then prevailing "mechanical philosophy" is a prime example of an external conceptual problem. As a general rule Laudan regarded conceptual problems as being more serious than empirical anomalies.

The rational appraisal of a research tradition is performed by assessing the "problem-solving effectiveness" of its component theories (Laudan 1977, 68):

> The overall problem-solving effectiveness of a theory is determined by assessing the number and importance of the empirical problems which the theory solves and deducting therefrom the number and importance of the anomalies and conceptual problems which the theory generates.

Laudan provided no calculus by which those assessments can be made. I suspect that no such calculus is possible (Newton-Smith 1981, 192–195).

One of Laudan's major contributions has been to distinguish among several different "cognitive stances" toward theories and traditions, particularly "pursuit" and "acceptance." It may, after all, make sense for a scientist to pursue a line of inquiry even though it is not at the moment the most successful approach. Like Kuhn and Lakatos, Laudan insisted that all such judgments are comparative. Thus, he said, it is rational to pursue a research tradition if its *rate* of problem-solving effectiveness is greater than that of its rivals. This may be the case even though the latest theory in one of the rival traditions has the highest overall problem-solving effectiveness.[22]

Like Lakatos, Laudan used his theory of scientific rationality to demarcate external from internal history. In doing so, he introduced an "arationality assumption" (1977, 202):

> The sociology of knowledge may step in to explain beliefs if and only if those beliefs cannot be explained in terms of their rational merits.

He elaborates as follows (1977, 202–3):

> The arationality assumption, we must stress, is a *methodological* principle, not a metaphysical doctrine. It does not assert that "whenever a belief can be explained by adequate reasons, then it could not have been socially caused"; it makes the weaker, programmatic proposal that "whenever a belief can be explained by adequate reasons, there is no need for, and little promise in, seeking out an alternative explanation in terms of social causes.

The arationality principle thus establishes a "division of labor between the historian of ideas and the sociologist of knowledge" (1977, 202), a division that clearly favors the historian of ideas.

METAMETHODOLOGY

As noted above, Lakatos introduced the term 'metamethodology' to describe his method for investigating the relative superiority of any proposed theory of scientific method. In *Progress and Its Problems* (1977, chap. 5) Laudan adopted a similar strategy—though differing in detail.

Laudan's metamethodological strategy was to seek first a set of "preferred pre-analytic intuitions about scientific rationality" (1977, 160). That is, looking at the history of science, we find (1977, 160):

> a subclass of cases of theory acceptance and theory rejection about which most scientifically educated persons have strong (and similar) intuitions. This class would include within it many (perhaps even all) of the following: (1) it was rational to accept Newtonian mechanics and to reject Aristotelian mechanics by, say, 1800; . . . ; (4) it was irrational after 1920 to believe that the chemical atom had no parts; . . . .

The next step was to apply the methodology (Laudan's theory) to these preferred cases (PIs) in order to determine the relative problem solving effectiveness of the traditions in question. A comparison of computed problem-solving effectiveness would tell us which tradition was in fact most progressive and thus which should have been accepted according to Laudan's methodology. "The degree of adequacy of any theory of scientific appraisal is proportional to how many of the PIs it can do justice to" (1977, 161).

The main trouble with any such metamethodology is that fails to avoid the relativism for which Kuhn was criticized. Beginning with *our* preanalytic intuitions, or *our* judgments of what constitutes great science, begs the interesting questions about relativism, or, in Lakatos's case, demarcation. Neither Lakatos nor Laudan would begin with the judgment that astrology or parapsychology count as great science, or are otherwise rationally acceptable.

Putting to one side the above fundamental objection, suppose that Laudan's methodology agreed with *all* our preferred intuitions. Could we then be confident that it is "a sound explication of what we mean by rationality" (1977, 161)? That seems not to follow. The most one could conclude is that Laudan had identified a highly reliable *symptom* of the basis for our preanalytic judgments of theory acceptance and theory rejection. For, suppose, contrary to Laudan, that our preanalytic judgments really are based on an assessment of the approximate truth of the theories in question, and that we take problem-solving effectiveness as our best evidence for approximate truth. Laudan's method of assessment would then yield the same judgments of acceptance and rejection but fail to capture the real basis of our judgments.

At its most successful, then, Laudan's metamethodology could tell us only that we had discovered a general description of situations we intuitively regard as clear cases of rational acceptance or pursuit. We might have correctly identified the descriptive component of the methodology, without capturing its normative force. To claim we had captured the normative component would require that we make the judgments we do *because* of considerations based on problem-solving effectiveness. In Kantian terms Laudan's metamethodology could at most show only that we are acting in accord with his methodology, not that we are acting out of regard for that methodology. It could not show that his methodology is actually embodied as a norm in our judgments.

This point is all the more pronounced if we consider not merely our own current preferred intuitions, but those of the historical actors in the episodes considered. Laudan did not attempt to show that actual scientists in historical contexts made the judgments they did because of considerations of problem-solving effectiveness. He was content to point out the correlation between their final judgments and our calculations of actual problem-solving effectiveness. That is scant evidence that such considerations were normatively in effect at the time. As a result, even metamethodological success would not justify using the methodology as a basis for explaining specific historical developments.

As far as Laudan himself is concerned, further arguments against the program of metamethodology are unnecessary, for he now rejects all "intuitionist" metamethodologies (Laudan 1986). His current approach to the evaluation of methodologies is more empirical.

The Reticulated Model

In *Science and Values* (1984a) Laudan began by noting that scientifc rationality has usually been taken as having a hierarchical structure: Theories are justified (relative to evidence) by reference to methodological rules, which in turn may be justified by appeal to more general aims. Justification flows only one way, from general aims to specific theories. Logical empiricists, including Popper, generally regarded the choice of aims as a matter of "convention." They were not troubled by charges of relativism because they assumed both their methodological rules and their aims were sufficiently abstract to be free of any specific content. Kuhn's global paradigms, by contrast, embraced theories, methods, and aims in a unified package. Thus, any choice among alternative paradigms must be non-rational because there is no neutral methodology or aim to which one could appeal. From most philosophical perspectives, conventionalism at the level of scientific theories is relativism.

Laudan maintains the components of the hierarchy but challenges the doctrine that rational influence flows only one way. Rather, he argues, general scientific aims can be challenged on the grounds that they cannot be achieved using available methods. And these grounds may themselves be established empirically. Similarly, specific theories may accord well with broad aims, such as generality, but, as a matter of empirical fact, not be capable of being authenticated by accepted methods. That, Laudan argues, is a rational ground for adopting different methods. Note, however, that these appeals to "rationality" are all of the instrumental, means-ends, variety (Laudan 1987).

Instead of a strict hierarchy, then, Laudan offers us an interacting triad of theory, method, and aims, as pictured in figure 2.2. Identifying a particular triad as a Kuhnian global paradigm, Laudan argues that we can get from one paradigm to another in a series of steps, each of which is instrumentally rational, and not, as Kuhn claimed, in one, nonrational leap of faith. The appearance of global paradigm change in historical cases is, Laudan argues, an artifact of too distant a historical perspective. Looking more closely, he claims, we see smaller, more piecemeal changes.

As several commentators have already noted (Doppelt 1986; Lugg 1986), it is not at all clear that changing paradigms a component at a time makes the resulting overall change any more rational than Kuhn's wholesale change. It seems possible to make rational changes within a complex whose adoption, overall, would not be rational. And so a theory of how rationally to change components within a paradigm is not enough for a gen-

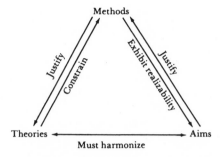

Figure 2.2. Laudan's reticulated model. Reproduced from Laudan (1984a, 63).

eral theory of scientific rationality. One needs also a way of rationally appraising the complex as a whole. And that Laudan has not yet provided.

THE REJECTION OF CONVERGENT REALISM

One point of agreement between logical empiricism and the historical tradition of Kuhn, Lakatos, and Laudan is the rejection of scientific realism. But only Laudan made that rejection explicit (Laudan 1981a; 1984a, chap. 5). Indeed, he regarded his arguments against scientific realism as an example of how one can use facts about accepted theory and method to argue against an axiology. Here we see the reticulated model of scientific rationality in action!

Interestingly Laudan has not denied that theoretical hypotheses may refer to real entities and processes and thus may be literally true or false. This he called semantic realism. What he has rejected is epistemological realism, roughly, the claim that many of our current scientific theories are in fact at least approximately true. Such claims, he has argued, are totally unfounded. His arguments for this position are various.

Perhaps Laudan's main concern has been to rebut the widespread view that realist claims are justified by the empirical success of science. This he has done by recalling the many theories that were empirically very successful for a time but later were rejected as fundamentally mistaken—phlogiston, the ether theory, and so on and on. Why, he asks, is our position vis-à-vis our current empirically successful theories any different? Indeed, the fact that most past theories have eventually been found wanting provides strong inductive evidence that the same fate will eventually overtake our current theories as well.

To such arguments realists have replied that our current theories are better approximations to the truth than previous theories. And we can expect that future theories will be better approximations still. So the fact that all our theories eventually prove mistaken in some respects is of no great consequence. What matters is that they were "approximately" true and that our approximations are getting better (Boyd 1973; Putnam 1975). This is the thesis Laudan calls "convergent realism."

Laudan regards the appeal to improved approximations as mere whistling in the dark. If the ether does not exist, there is no coherent sense of approximation according to which the ether theory of electromagnetic radiation can be said to have been even approximately true.[23] Many philosophers of science remain unconvinced by Laudan's arguments, but no realistic theory of science can be viable if it fails to account for the historical evidence Laudan presents.

The influence of Kuhn on Lakatos and Laudan (as well as on many other post-Kuhnian philosophers of science) is obvious. Less obvious is how much these philosophers share with their logical empiricist predecessors, and how little in the way of actual doctrine, as opposed to a general focus on the historical development of science, they share with Kuhn.

The terms 'research programs' or 'traditions' are obviously derived from Kuhn's notion of paradigm-governed normal science. But the sense of paradigm adopted is exclusively Kuhn's global sense, which includes theory, method, aims, metaphysics, and so on. Thus, both Lakatos's research programs and Laudan's research traditions are identified with sets of statements, that is, empirical laws, methodological rules, and so forth. The categories in which both Lakatos and Laudan formulate their theories are all of this abstract type. Individual scientists as real people are as absent from their theories as they were from the analyses of the logical empiricists.

For Kuhn, as I understand him, it is the paradigm as *exemplar* that is primary. What counts are the judgments of individual scientists that a particular solution provides a model for further research. The abstract formulation of laws and methodological rules plays much less of a role. Post-Kuhnian philosophers of science have largely ignored, or rejected, these aspects of Kuhn's account. But it is just these aspects that seem to me most suggestive and promising for further development.

Constructive Empiricism

The Kuhnian revolution in the philosophy of science, like most revolutions, was neither instantaneous nor all-inclusive. Its full influence took more than a decade to unfold, and even by then some areas of the philosophy of science had barely felt its impact. Work in the foundations of physics and in the study of probability and induction, for example, continued on its own course for a generation. Thus is it possible for one of the most influential of recent books in the philosophy of science, van Fraassen's *The Scientific Image* (1980), to exhibit little trace of Kuhn's thought. It is a masterful defense of empiricism that eliminates the positivist excesses of its predecessors.

Van Fraassen rejected the logical empiricist view of theories as interpreted formal systems. He rejected any attempt to construct an inductive logic. And he rejected the logical analysis of scientific explanation. Nevertheless, van Fraassen's positive account of science upholds the deflationary anti-realism that inspired logical empiricism. Here I shall consider only his view of theories and his anti-realism since these bear directly on themes in my own account.[24]

THE "SEMANTIC" VIEW OF THEORIES

Logical empiricists portrayed scientific theories as uninterpreted formal systems that house empirical rules of interpretation. A formal system itself is a set of statements ("axioms") formulated in an explicitly characterized, formal language. Elements of the formal language were characterized solely in terms of their syntactic structure. *Fx,* for example, is a one-place predicate function; *Rxy,* a two-place relational function; and *Fa&Fb,* a conjunction of two singular statements, where *a* and *b* are the names of particular objects. Important logical relationships, such as valid inference, were also characterized purely syntactically. Thus, from *Fa&Fb* one can validly infer *Fa* simply because of the formal, syntactic structure of these two statement forms.

Semantic notions such as meaning and truth entered the logical empiricist picture of theories only indirectly by means of the interpretative rules employed. A minimal sort of interpretation serves to introduce the important semantic notions of "structure" and "model." Imagine some axioms formulated in a simple first-order language, *L*. A *structure* for *L* is a set of objects, *O,* and a function that assigns subsets of *O* to one-place predicates of *L,* ordered pairs of objects to two-place relations, and so on. A *model* of a theory *T,* expressed as axioms in *L,* is any structure in which the axioms of *T* are true. The concept of a model, being defined in terms of truth, is therefore a semantic, as opposed to a syntactic, concept.

Whether any given structure is indeed a model of *T* requires independent determinations of the truth of the axioms. This may be done mathematically if *O* is a set of mathematical entities, such as integers, or empirically if *O* is a set of physical objects, such as planets. In any case, models are *nonlinguistic* entities—sets of objects, not sets of statements.

For any theory expressed as a set of axioms in a formal language, there is, therefore, a set of structures that are models of that theory. This way of presenting models makes them appear to be derivative upon the language in which the axioms are formulated. But this appearance is merely an artifact of our having started with axioms formulated in a particular language and then having looked for structures in which the axioms turn out to be true. Since models are nonlinguistic entities, they can be characterized in many different ways, using many different languages. This makes it possible to identify a theory not with any particular linguistic formulation, but with the *set of models* that would be picked out by all the different possible linguistic formulations.

Van Fraassen (1980, 44) expresses the contrast between the syntactic and the semantic approaches as follows:

> The syntactic picture of a theory identifies it with a body of theorems, stated in one particular language chosen for the expression of that theory. This should be contrasted with the alternative of presenting a theory in the first instance by identifying a class of structures as its models. In this second, semantic, approach the language used to express the theory is neither basic nor unique; the same class of structures could well be described in radically different ways, each with its own limitations. The models occupy centre stage.

Van Fraassen's semantic conception of theories liberates the philosophical study of science from the linguistic shackles of its logical empiricist predecessor.

Logical empiricists were well aware that scientists themselves do not present their theories in the form of interpreted formal systems. Thus, the philosophical analysis of any science required a prior "rational reconstruction" of the content of that science in an appropriate formal language. Focusing on models allows one to bypass that step. One is therefore free to use any sufficiently rich language, including the language of the scientists themselves, in talking about the content of any science.

Empirical Adequacy

Logical empiricists divided the vocabulary of a theory into "observational terms" and "theoretical terms." The former meaningfully referred to aspects of the real world; the latter derived their meaning from their logical connections with observational terms and with one another. The result was instrumentalism with regard to the supposed "theoretical entities" of modern science.

Van Fraassen reinstated a subtle, nonlinguistic version of this distinction between the observable and the theoretical. The result is not, strictly speaking, instrumentalism. But it is a form of anti-realism. His capsule formulation of his "New Picture of Theories" (1980, 64) reads:

> To present a theory is to specify a family of structures, its *models;* and secondly, to specify certain parts of those models (the *empirical substructures*) as candidates for the direct representation of observable phenomena.

In short, models have two parts, an empirical substructure and a theoretical superstructure (the latter being my term, not van Fraassen's).

Van Fraassen did not deny that the theoretical superstructure may represent features of the real world. In logical empiricist categories theoretical terms may meaningfully refer to, and statements employing such terms may literally be true of, the world. Nevertheless, he did deny that science need be concerned with any such match between the theoretical superstructure and the world. It is sufficient for the purposes of science that there be a match between the empirical substructure and the observable phenomena. A theory (set of models) exhibiting this match is called "empirically adequate." Constructive empiricism holds that science aspires not to full, literal truth but merely to empirical adequacy.

Sometimes van Fraassen (1980, 69) framed the distinction between realism and constructive empiricism in terms of "belief" and "acceptance."

> With this new picture of theories in mind, we can distinguish between two epistemic attitudes we can take up toward a theory. We can assert it to be true (i.e. to have a model which is a faithful replica, in all detail, of our world), and call for belief; or we can simply assert its empirical adequacy, calling for acceptance as such. In either case we stick our necks out; empirical adequacy goes far beyond what we can know at any given time. . . . Nevertheless, there is a difference: the assertion of empirical adequacy is a great deal weaker than the assertion of truth, and the restraint to acceptance delivers us from metaphysics.

Deliverance from metaphysics was a major statement in the creed of logical empiricism.

A fuller statement of van Fraassen's own credo appears near the end of the book (1980, 202–3):

To be an empiricist is to withhold belief in anything that goes beyond the actual, observable phenomena, and to recognize no objective modality in nature. To develop an empiricist account of science is to depict it as involving a search for truth only about the empirical world, about what is actual and observable. Since scientific activity is an enormously rich and complex cultural phenomenon, this account of science must be accompanied by auxiliary theories about scientific explanation, conceptual commitment, modal language, and much else. But it must involve throughout a resolute rejection of the demand for an explanation of the regularities in the observable course of nature, by means of truths concerning a reality beyond what is actual and observable, as a demand which plays no role in the scientific enterprise.

Constructive empiricism, then, pictures science as involving the construction of models and the testing of those models against observable phenomena to judge their empirical adequacy.

It sometimes seems as if this picture is presented merely as a philosophically *possible* account of science. At other times, however, van Fraassen (1980, 73) is explicit in asserting at least the empirical adequacy of his own theory of science:

There is also a positive argument for constructive empiricism—it makes better sense of science, and of scientific activity, than realism does.

I shall not here raise objections to this claim. In later chapters I shall myself adopt something like the "constructive" part of constructive empiricism, while rejecting the accompanying empiricism. The result will be called *constructive realism*.

The Strong Program

Kuhn's work had little immediate impact on the Mertonian school of American sociologists of science. Rather, his influence on the sociology of science appeared first in Great Britain, and even then not so much directly within sociology as around the Science Studies Unit at the University of Edinburgh. Formed in the late 1960s by David Edge, a scientist turned sociologist of science, the Science Studies Unit assembled a number of historians and sociologists including Barry Barnes, David Bloor, Stephen Shapin, and Donald MacKenzie. By the early 1980s members of this group, together with other British and European allies, were putting together anthologies of papers written from their perspective—a sure sign of a movement's reaching maturity (or at least reaching *for* maturity).[25]

Kuhn, of course, was not the only source of inspiration for members of the Edinburgh school. In some cases he was not even a major source. Inspiration came also from the general sociology of knowledge as found in Durkheim (1915) and Mannheim (1952), the philosophy of Wittgenstein (1953) and followers of Wittgenstein such as Peter Winch (1958), the anthropology of Mary Douglas (1970), the "critical theory" of Jurgen Habermas (1972), and the ethnomethodology of Harold Garfinkel (1967). It was David Bloor (1976, 4–5) who baptized these related strands. He called them "the Strong Program in the sociology of scientific knowledge."

> The sociology of scientific knowledge should adhere to the following four tenets:
> 1. It would be causal, that is, concerned with the conditions which bring about belief or states of knowledge. Naturally there will be other types of causes apart from social ones which will cooperate in bringing about belief.
> 2. It would be impartial with respect to truth and falsity, rationality or irrationality, success or failure. Both sides of these dichotomies will require explanation.
> 3. It would be symmetrical in its style of explanation. The same types of cause would explain, say, true and false beliefs.
> 4. It would be reflexive. In principle its patterns of explanation would have to be applicable to sociology itself. . . .
> These four tenets, of causality, impartiality, symmetry and reflexivity define what will be called the strong programme in the sociology of knowledge.

From these four tenets we can infer some major general features of an Edinburgh-style theory of science.

First, it is a theory that deals directly with the *content* of science. It is concerned not merely with the "social structure" in which science operates, or even with the conditions that foster or inhibit science. It is concerned with "the conditions which bring about belief or states of knowledge."

Second, it is militantly, and not merely implicitly, *naturalistic*. "The sociologist," wrote Bloor (1976, 4), "is concerned with knowledge, including scientific knowledge, purely as a natural phenomenon." The strong program thus explicitly denies any fundamental distinction between "reasons" and "causes." It is part of the Edinburgh conception of natural science that it is fundamentally causal. That is why these explanations are to be causal. According to the strong program, then, it is never sufficient to explain a scientific belief by saying that it is true, or that it follows logically from the evidence, or that it is rational. A causal explanation is required.

Bloor explicitly rejected Lakatos's demarcation between rational history and irrationality. He later (Bloor 1981) also rejected Laudan's arationality principle. In any such view "the sociology of knowledge is confined to the sociology of error" (1976, 8). All the sociologist can do is "close the gap between rationality and actuality" (1976, 7). According to the strong program, on the other hand, a sociological explanation of scientific belief can in principle be as complete as any explanation in any other science. The strong program thus leaves even less for rational historians like Lakatos and Laudan than those historians left for sociologists.

Third, the naturalism of the strong program extends to the sociology of science itself. The sociology of science is a science like any other and is thus part of its own subject matter. Merton long ago noted the reflexivity of the sociology of science. What is new in the strong program is the extension of reflexivity to include the "beliefs," that is, the theories, of the sociology of science, and not merely its social structure.

Unlike a Mertonian sociology of science, which can be seen as a complement to either logical empiricism or later philosophical theories of science, an Edinburgh-style cognitive sociology of science complements Kuhn's intuitive sociology. It, like Kuhn's theory of science, is in direct conflict with all philosophical theories that seek to distinguish logic or rationality from psychology or sociology.

THE SOCIOLOGICAL HISTORY OF SCIENCE

The Edinburgh school has not contented itself with programmatic statements. Indeed, its advocates prefer to point to actual case studies. As Shapin (1982, 157–58) recently wrote, "One can either debate the possibility of the sociology of scientific knowledge, or one can do it." He then went on to discuss "the many empirical successes of practical sociological approaches to scientific knowledge." His numerous examples provide a good survey of the range of causal factors invoked to explain scientific belief.

The general form of explanation in fact favored by the Edinburgh school appeals to the *interests* of scientists in explaining their scientific beliefs. The main difference in the studies Shapin discussed is the *type* and relative importance of the interests invoked.

Following Habermas (1972) it is assumed that all scientific work since the seventeenth century has involved, to some extent, an interest in prediction and control. But, it is argued, this interest, no matter how strong, even when coupled with observational or experimental data, and no matter how extensive, does not uniquely determine scientific beliefs. Sometimes this point is argued with explicit reference to the philosophical literature on the "underdetermination" of theory by data. The conclusion, of course, is that an adequate explanation of scientific beliefs requires something more, which in practice is the influence of other interests beyond those of prediction and control. Our attention is thus directed to those other interests.

At the widest scope are interests associated with social class or social position. Here Shapin cited his own studies of phrenology in late eighteenth century Edinburgh (Shapin 1975, 1979). Belief in the doctrines of phrenology, including quite specific neurophysiological findings, was, we are told, strongly associated with, and served the social interests of, "rising bourgeois groups." Similarly opposition to phrenology was concentrated in, and served the social interests of, the "traditional elites." With an allusion to underdetermination, Shapin concluded (1982, 4):

> Reality seems capable of sustaining more than one account given of it, depending upon the goals of those who engage with it; and in this instance at least those goals included considerations in the wider society such as the redistribution of rights and resources among social classes.

Shapin drew similar conclusions from Farley and Geison's (1974) study of the mid-nineteenth century debate over spontaneous generation between Pouchet and Pasteur and from MacKenzie's (1978, 1981) study of the dispute over the correlation coefficient between G. U. Yule and Karl Pearson.

As an intermediate case Shapin cited John Lankford's (1981) study of a late nineteenth century controversy between British amateur astronomers and American professionals over planetary observations such as the "canals" on Mars. The American position, Lankford claimed, was as much determined by the professional interest of the American astronomers in upholding the superiority of their instruments, and thus their status as professionals, as by what was observed. In this case it was professional interests, not wider social interests, that mattered most.

At the other extreme Shapin cited many studies of contemporary science in which the interests attach to the professional status or reputation of *individual* scientists. Scientists, it is claimed, tend to believe what they do at least in part because their individual reputations are at stake.

Sociologists who advocate strong program explanations generally do more than draw a simple causal connection between interests and beliefs. They often try to say something about how the connection works. Here the primary notion is that of *negotiation*. Scientists, it is said, negotiate on behalf of their interests. The result of their negotiations is said to be a "socially constructed" body of scientific knowledge. Unfortunately talk about negotiation is generally little more than a gloss. Not much is said about the actual *processes* of negotiation. so the connection between interests and beliefs remains somewhat vague.[26]

The Empirical Relativist Program

The first reaction of many philosophers to the strong program is that it is caught in a hopeless, if not outright self-refuting, epistemological relativism. Indeed, the sociologists' professional interests in denying the claims of rational historiography are all too apparent. Advocates of the strong program and their allies have been aware of this charge from the start and have tried, defensively at first, but then more confidently, to meet it.

In the epilogue to *Scientific Knowledge and Sociological Theory,* Barry Barnes (1974, 156) wrote:

> Why, it will be said, if the preceding account makes no claims to being the best account we have in a fully objective sense, should anybody accord it credibility? . . . The answer is easily given; this whole volume is crammed with proffered reasons why its main tenets should be accepted; its justification lies within itself. . . . The point is that this can only hope to count as a justification within particular forms of culture. What the range of forms might be I am far from sure; hopefully what has been said will carry weight among some of those engaged in sociological activity and related fields. Perhaps it is sufficiently congruent with the common sense of our general culture to be thought plausible or even compelling over a wider field. But just as there is always an uncomfortable gap between any large integer and infinity, so will there always be a gap between justification according to particular conventions, and justification in an unqualified sense.

Bloor's remarks in *Knowledge and Social Imagery* are similar. Later we find both Barnes and Bloor (1982: 21–22) taking a more aggressive stance:

> Far from being a threat to the scientific understanding of forms of knowledge, relativism is required by it. Our claim is that relativism is essential to all those disciplines such as anthropology, sociology, the history of institutions and ideas, and even cognitive psychology, which account for the diversity of systems of knowledge, their distribution and the manner of their change. It is those

who oppose relativism, and who grant certain forms of knowledge a privileged status, who pose the real threat to a scientific understanding of knowledge and cognition.

Here relativism is seen not as something to be avoided, but embraced.

The most outspoken defender of a relativist perspective in the sociology of science has been Harry Collins, who advocates an "empirical program of relativism" (Collins and Cox 1976; Collins and Pinch 1982; Collins 1981a, 1981b, 1982). Collins's official position has been that relativism is a "favourable heuristic for the sociology of science" (Collins and Cox 1976, 423). Pursuing this heuristic will extend the boundaries of sociological explanations of science to their maximum (Collins and Cox 1976, 438–39):

> If writers wish to soothe themselves and others by endorsing an "objective" world, they should make clear exactly where they think it must intrude in accounts of scientific practice—preferably with examples. If "nature" is unambiguous, then to discover its point of entry into scientific investigation would be to solve a modern epistemological problem. The approach we favour is to push the relativistic heuristic as far as possible: where it can go no further, "nature" intrudes.

The heuristic stance, however, is difficult to maintain. Thus, we find Collins (1981a, 3) embracing:

> an explicit relativism in which the natural world has a small or non-existent role in the construction of scientific knowledge.

The relativist program, he claims (1981b, 6–7) implies that:

> in one set of social circumstances "correct scientific method" applied to a problem would precipitate result p whereas in another set of social circumstances "correct scientific method" applied to the same problem would precipitate result q, where, perhaps, q, implies not-p.

This does not sound like a merely "heuristic" stance.

Such a militant relativism has, of course, drawn fire from philosophers and other sources (Laudan 1981b, 1984b; Hollis and Lukes 1982). Much of the debate, however, seems to me misguided. If, as Collins often complains, relativism is identified with "irrationalism," then of course arguing for relativism is self-defeating. But that line misses the point. The real issue, as I see it, is the extent to which, and by what means, nature constrains scientific theorizing.

It is, however, not only the critics of the strong program who have muddied the waters. Its advocates repeatedly cite Evans-Pritchard's (1937) studies of witchcraft among the Azande, implying that modern science is just our own version of witchcraft. Now, it is one thing to reject explanations of scientific beliefs that appeal to the truth or rationality of these beliefs. It is quite another to refuse to acknowledge that our current scientific beliefs are in some important sense better than those of the past. That how we came to embrace our current scientific beliefs is not to be explained by appeal to their truth or to our rationality surely does not imply, for example, that we do not in fact now know much more about biology and chemistry than was known in the eighteenth century. An empirical theory of science need not deny such obvious facts. Its task is to *explain* them.

Laboratory Studies

Those working within the strong program generally take the unit of analysis to be a research specialty. Typically one looks at specialties over a span of time in which there is active debate among the scientists themselves over the relative merits of two or more theories. The assumption is that times of controversy are particularly revealing of the mechanisms by which science operates. Such controversies typically involve ten to a hundred scientists operating in several different laboratories, perhaps even in several different countries.

Within the strong program people talk about "the social construction of scientific knowledge" (MacKenzie 1981). To some of their colleagues in sociology, however, the claim of social construction is diluted or unsubstantiated because the research specialty is too coarse a unit of analysis to reveal the detailed, microsociological processes through which knowledge can be constructed. For that, they claim, one must look at day-to-day research in the laboratory.

The Constructivist Program

The first, and still most influential, laboratory study, Bruno Latour and Stephen Woolgar's *Laboratory Life* (1979) (subtitled "The Social Construction of Scientific Facts"), was based on Latour's two years of field work at the Salk Institute in 1975–77. The authors' approach to the subject is dramatically revealed in the following passage (1979, 128–29). The substances referred to are various "releasing factors" studied by neuroendocrinologists at the Salk Institute during Latour's visit. "Inscriptions" are any recorded symbols, including sentences written by scientists and graphs drawn by computers.

> We have attempted to avoid using terms which would change the nature of the issues under discussion. Thus, in emphasizing the process whereby substances are *constructed,* we have tried to avoid descriptions of the bioassays which take as unproblematic relationships between signs and things signified. Despite the fact that our scientists held the belief that the inscriptions could be representations or indicators of some entity with an independent existence "out there," we have argued that such entities were constituted solely through the use of these inscriptions. It is not simply that differences between curves indicate the presence of a substance; rather the substance is identical with perceived differences between curves. In order to stress this point, we have eschewed the use of expressions such as "the substance was discovered by using a bioassay" or "the object was found as a result of identifying differences between two peaks." To employ such expressions would be to convey the misleading impression that the presence of certain objects was a pregiven and that such objects merely awaited the timely revelation of their existence by scientists. By contrast, we do not conceive of scientists using various strategies as pulling back the curtain on pregiven, but hitherto concealed, truths. Rather, objects (in this case, substances) are constituted through the artful creativity of scientists.

The contrast between "construction" and "representation" is explicit.

The most consistent champion of laboratory studies has been Karin Knorr-Cetina, whose book *The Manufacture of Knowledge* (subtitled "An Essay on the Constructivist and Contextual Nature of Science") appeared in 1981. She too emphasized the contrast between "description" and "construction," or between "fact" and "artifact." In a more recent summary of her views, she wrote:

> The constructivist interpretation is opposed to the conception of scientific investigation as descriptive, a conception which locates the problem of facticity in the relation between the products of science and an external nature. In contrast, the constructivist interpretation considers the products of science as first and foremost the result of a process of (reflexive) fabrication. Accordingly, the study of scientific knowledge is primarily seen to involve an investigation of how scientific objects are produced in the laboratory rather than a study of how facts are preserved in scientific statements about nature. (1983, 118–19)

It is the thrust of the constructivist conception to conceive of scientific reality as progressively emerging out of indeterminancy and (self-referential) constructive operations, without assuming it to match any pre-existing order of the real. (1983, 135)

While the critical reader may question whether it is "knowledge" or "the facts" that are being socially negotiated, it is undeniable that these works of Latour and Woolgar and of Knorr-Cetina capture the texture of day-to-day research in a way that few other works, be they sociological, historical, or philosophical, have ever done. Still, for anyone trained in the natural sciences or in an analytic philosophy of science, constructivism sounds wildly implausible. Could it be that there is a sufficient disparity in intellectual background that champions of constructivist programs are not really saying what, to someone with a different background, they seem to be saying?

THE SOCIAL CONSTRUCTION OF SOCIAL REALITY

The constructivist program for the study of science has roots in the European sociological tradition going back to Scheler, Durkheim and Mannheim, and before that Marx and Nietzsche. It also has obvious ties to phenomenology. But we need not plumb these depths to get a better fix on the constructivist program. It is sufficient, I think, to consult a book that is often cited (though seldom discussed) in the recent constructivist literature: Berger and Luckmann's *The Social Construction of Reality* (1966). The connection to the older European tradition and to phenomenology is through Luckmann by way of Alfred Schutz.

The revealing fact is that Burger and Luckmann's book is about the social construction of *social* reality. The two main chapters, in addition to the introduction and conclusion, are titled "Society as Objective Reality" and "Society as Subjective Reality." The authors explicitly declined to pursue questions about the role of "modern science" in society (1966, 112). And in considering relationships between biology and society, they explicitly noted that some socially conceivable projects, such as legislating that men should bear children, "would founder on the hard facts of human biology" (1966, 181). Yes, "hard facts"!

The idea that social facts might be socially constructed with no necessary relationship with a preexisting reality is not difficult to understand, or even to accept. Examples abound. Facts about kinship relations, for example, whether a first cousin is a possible marriage partner, are socially constructed—and vary from culture to culture. Facts about corporations, such as their tax liabilities, are obviously social constructs. Indeed, taking seriously the social construction of social facts leads to the curious and somewhat ironic suggestion that *scientists* themselves are likely candidates for socially constructed entities. No one is born a scientist, though some people are born with capabilities that make for a good scientist. Thinking historically, one could make a good case that scientists were socially invented sometime after 1700, probably after 1800.

Do advocates of the constructivist program in the sociology of science really mean to claim that scientific objects and facts have a similar status to social objects and facts? It seems to me the answer must be yes. That is what Knorr-Cetina meant when she said (1983, 136) that "this known world is a cultural object, a world identified and embodied in our language and our practice." The world scientists construct is a cultural object in the same sense that cousins, corporations, and scientists are cultural objects.

Further evidence for taking at face value the claim that scientific constructions are like other social constructions comes from the fact that constructivists typically argue that there is no fundamental difference between the social sciences and the natural sciences. The final chapters of both *Laboratory Life* and *The Manufacture of Knowledge* argue this thesis. And indeed, the thesis follows from the claim that scientific facts are socially constructed in the way that social facts are. On this understanding, the subject matter of both the natural and the social sciences has fundamentally the same status.

REALITY: SOCIAL AND NATURAL

The general idea of "social construction," then, can be accepted for many aspects of *social* reality. But this, by itself, provides no evidence that *natural* reality is similarly constructed. That conclusion requires independent evidence. Of course, the practitioners of laboratory studies claim to have presented such evidence. In chapter 5 I shall consider some of this evidence and confront it with a laboratory study of my own.

The Sociological Analysis of Scientists' Discourse

Whether focusing on broad social interests or social interaction in the laboratory, most recent sociologists of science have assumed that the objective of sociological investigation is to explain the beliefs and actions of scientists. This assumption is now being challenged by Michael Mulkay and his collaborators (Mulkay 1974, 1979, 1981; Gilbert and Mulkay 1981, 1982; Gilbert 1980; Mulkay, Potter, and Yearley 1983).

A passage from the introductory chapter of Gilbert and Mulkay's *Opening Pandora's Box* (1984, 2) outlines their perspective:

Most sociological analyses are dominated by the authorial voice of the sociologist. Participants are allowed to speak through the author's text only when they appear to endorse his story. Most sociological research reports are, in this sense, univocal. We believe that this form of presentation grossly misrepresents the participants' discourse. This is not only because different actors often tell radically different stories, but also because each actor has many different voices. . . .

The goal of constructing definitive analysts' accounts of scientists' actions and beliefs is possibly unattainable in principle, and certainly unattainable in practice as long as we have no systematic understanding of the social production of scientists' discourse. Sociologists, historians and philosophers have been able to document and make plausible so many divergent analyses of science (and continually undermine each other's claims) because scientists, the active creators of analysts' evidence, themselves engage in so many kinds of discourse. Thus we recommend that analysts should no longer seek to force scientists' diverse discourse into one "authoritative" account of their own. Instead of assuming that there is only one truly accurate version of participants' action and belief which can, sooner or later, be pieced together, analysts need to become more sensitive to interpretative variability among participants and to seek to understand why so many different versions of events can be produced.

The "social world of science," they conclude (1984, 188), is to be approached as a "multiple reality."

Opening Pandora's Box is a mostly retrospective study of 20 years of biochemical research on oxidative phosphorylation. In addition to examining published and unpublished materials, Gilbert and Mulkay conducted two- to three-hour interviews with 34 biochemists, roughly half of the people who had been active in the field. Despite occasional qualifications these "analysts" rejected any attempt to show that the general consensus in the field in 1980 was in any objective sense an advance beyond that which existed in 1960. Rather, the most the analyst of science can do, they insisted, is look for regularities in how scientists employ interpretative resources in representing these scientific developments to themselves and others.

Gilbert and Mulkay attempted to show that scientists employ at least two different linguistic repertoires: an *empiricist* repertoire, in which the "correct" theory is seen as uniquely determined by experimental data; and a *contingent* repertoire, in which beliefs and actions are explained by reference to a variety of social, professional, and personal factors. Scientists,

Gilbert and Mulkay argued, typically use the empiricist repertoire to explain their own, "correct," beliefs, whereas they employ the contingent repertoire to account for the fact that other scientists with access to the same data hold opposing views. Scientists portray themselves as guided soley by the experimental data, while charging that those who hold opposing views are "dogmatic," or "have too much invested in a theory to give it up," and so on.

Anyone who has poured over transcripts from interviews with scientists will recognize the phenomena that led Gilbert and Mulkay to postulate two different repertoires. It is indeed difficult to construct a single, coherent account even from an interview with just one scientist. The problem Gilbert and Mulkay raised is a real problem. And sensitivity to the various interpretative resources used by scientists is important. Yet the resort to "multiple realities," rather than merely multiple *accounts,* or at most multiple *social* realities, seems an overreaction to an admittedly difficult situation.

Ironically the retreat from ascriptions of belief to the analysis of discourse is more in keeping with philosophical empiricism, or with behaviorism in psychology, than with cognitive movements in sociology and anthropology. It was behaviorism, after all, that rejected inferences from "verbal behavior" to internal, representational states. Contemporary cognitive scientists, by contrast, regularly ascribe cognitive states, some considerably more complex than ordinary belief states, to human subjects. Perhaps cognitive scientists are sometimes overconfident in their ability to do this well. And perhaps scientists in their own labs are more difficult subjects than the typical subject in a psychologist's lab. But restricting the study of science to analyzing patterns in scientists' discourse seems a counsel of despair.

3

Models and Theories

One of the primary means by which scientists represent the world is through the use of theories. If it is true that scientists produce knowledge, an important part of that knowledge is *theoretical* knowledge. Any account of science must face questions like "What are theories?" and "How do theories function in various scientific activities?"

Among students of the scientific life only the logical empiricists and some more recent analytic philosophers of science have developed detailed, general accounts of what theories might be. Other historians and philosophers of science, as well as sociologists of science, have generally been content to speak of "concepts," "beliefs," "hypotheses," or, indeed, "theories" without providing a further account of what these might be—an account that employs resources beyond those of folk psychology. Recent work in the cognitive sciences provides the inspiration, and some resources, for going further in directions already indicated by analytic philosophers such as Suppes (1967, 1969), van Fraassen (1970, 1980), and Suppe (1972, 1973).

I shall not, however, approach the question of what scientific theories might be by looking first at what cognitive scientists have to say about how, in general, humans represent their world. I shall begin, instead, with *scientific* representations themselves, keeping in mind that scientists, after all, are only human. The representations scientists construct cannot be too radically different in nature from those employed by humans in general.

The Science Textbook

The transmission of scientific knowledge has now become quite uniform. It relies heavily on the advanced *textbook*. Until beginning dissertation research most scientists in most fields learn what theory they know from textbooks—in conjunction with lectures that also follow a textbook format. If we wish to learn what a theory is from the standpoint of scientists who use that theory, one way to proceed is by examining the textbooks from which they learned most of what they know about that theory. This is obviously not the only way to learn what theories are to scientists themselves. I would not even claim that it is necessarily the best way to proceed. But it is a good way, and one that is accessible to someone like me whose training has been primarily in science and philosophy.

Historians of science, such as Kuhn, have charged that textbooks distort the history of the subject to the point of providing historically false accounts of how various theories came to be accepted. Although there is no doubt much truth in this charge, it is of no consequence here. The task here is not to reconstruct the historical development of any science but simply to describe, in general terms, the character of a theory as it is understood by contemporary scientists.

Many philosophers of science would also object that textbooks provide a poor place to start. Textbooks are notoriously vague about fundamentals and by no means consistent one with another. Better to look at the work of philosophers of science and the few scientists concerned with the "foundations" of their subject. It is clear from the preceding chapters, however, that this approach will not do. Philosophers pursuing what they call foundational studies *begin* with a conception of what a theory ought to look like and seek to *reconstruct* theories in that mold. The question whether their conception matches that actually employed in science is begged from the start. Moreover, these foundational studies have had almost no impact on the way science is taught, learned, or understood by scientists. Is one to say that, on the whole, authors of widely used scientific textbooks do not really understand their subject? If so, then most scientists do not really understand their subject, because that is where they learned most of what they know. Such a view is uncomfortably arrogant, if not simply absurd.[1]

In this chapter I shall focus on classical mechanics. The choice is to some extent opportunistic. Classical mechanics is among the theories I know best. Moreover, I first learned it as an aspiring physicist, at both the undergraduate and the graduate levels. Thus, my own understanding of the theory rests not simply on reading the texts, but also on having experienced the acculturation process that is part of becoming a scientist.

The choice of classical mechanics is not solely opportunistic, however. A number of facts make it a good place to begin investigating the nature of scientific theories. One is that classical mechanics has been discussed by almost everyone who has written on the nature of theories. If one is to enter this debate with an alternative account, one is obliged to explain why the alternative does or does not apply to classical mechanics.

Second, classical mechanics is typical of a wide range of theories, particularly, though not exclusively, in the physical sciences. Moreover, there is a ready historical explanation of why mechanics is indeed typical. Because it was the first modern scientific theory, mechanics has been the prototype for other theories both within physics and beyond. In particular, the mathematical techniques created in the process of developing the science of mechanics, such as the calculus, have been applied far beyond the confines of physics. In short, mechanics must be fairly typical of many scientific theories simply because so many other theories have been explicitly modeled on it.

A third reason for beginning with classical mechanics is that the education of many scientists still includes physics, and the serious study of physics begins with classical mechanics. For a great many contemporary scientists, then, classical mechanics provided their first serious exposure to a scientific theory. It cannot have failed to influence their general conception of what constitutes a genuinely scientific theory.

Nonetheless, in spite of these reasons a serious question remains as to just how widely the account to be developed might be applied. I believe the account employs categories not only general enough to be widely applicable, but also specific enough to carry real content. But the proof must eventually be found in actual applications.

The Organization of a Mechanics Text

Few truly elementary texts exist in mechanics. Elementary treatments typically appear in more general texts that also cover several other areas of physics, such as electricity and magnetism and atomic physics. The broad range of texts specializing in mechanics fall in the category of intermediate to advanced, the difference being primarily in the sophistication of the mathematical framework employed. In addition, there are truly professional treatises devoted either to advanced mathematical formulations or to advanced applications. The focus of the present inquiry should definitely be on the intermediate to advanced category. That category includes most of the texts studied by the majority of undergraduate majors and graduate students—the next generation of professional physicists.[2]

Some texts begin with a chapter or two of mathematical preliminaries. These books seem to be primarily the less advanced texts in which the author wishes to use the vector calculus or some other mathematical methods that one cannot always assume the student knows. Texts that simply employ the calculus tend to assume the student has already learned it.

A few of the generally less advanced texts devote a preliminary chapter to kinematics, the description of motion using such basic notions as position, velocity, and acceleration. Some authors use this opportunity to introduce vector notation and polar coordinates.

The first substantive chapter of any textbook almost invariably includes a statement of Newton's laws. It is typical to focus on the second law, but many present all three. The general descriptions of the laws are various. One text, claiming to follow Mach, calls the first and second laws definitions and ascribes all the physical content to the third law. Another (Goldstein 1959, 1) ascribes all the "essential physics" to the second law, which the author claims "may be considered equivalently as a fundamental postulate or as a definition of force and mass." A third author considers the possibility that Newton's laws define force, while the special force laws are empirical laws. After considering the reverse position as well, he concludes (Symon 1953, 8):

> Probably the best plan, the most flexible at least, is to take force as a primitive concept of our theory, perhaps defined operationally in terms of measurements with a spring balance. Then Newton's laws are laws, and so are the laws of theories of special forces like gravitation and electromagnetism.

To the philosopher of science this proposition may seem hopelessly confused, reinforcing the prejudice that writers of textbooks do not really know what they are doing.

At this point, however, trying to get straight which statements are postulates and which are definitions would be a diversion. The real question is why it does not seem to matter to the learning and doing of physics which, if any, of Newton's laws one calls postulates and which definitions? Could it be that this distinction does not function here in the way philosophers and logicians would have us believe?

Let us proceed, following the path that focuses on the second law: Force equals mass times acceleration, or:

$$(3.1) \qquad F = ma = m \, d^2x/dt^2.$$

The force acting on a body, then, is equal to its mass times the *second* derivative of its position with respect to time. The second law, therefore, has the mathematical form of a second-order differential equation.

Few texts even pause to question the assumption that position in space is twice differentiable or that the notions of velocity and acceleration at an instant make sense—ideas that much perplexed Newton and his contemporaries. These notions are today largely taken for granted.

Now an important fact emerges. From the second law, or from all three, nothing much of interest follows. If the laws of motion by themselves are taken as the axioms of an axiomatic system, it is a rather uninteresting one. Nor do any texts I know proceed to deduce theorems from these "axioms." Rather, they typically proceed to a series of chapters based on different assumptions about the form of the force function. The grouping and sequencing of these chapters vary. Nevertheless, it is *force functions* that provide the chief organizing principle in most classical mechanics textbooks.

A typical progression would be to proceed first to treat motion in one dimension. Within that category one might progress from uniform forces, to forces as a function of position only, to forces that are a function of both position and velocity, and finally to forces that are a function of position, velocity, and time.

Each of these possible force functions is illustrated with one or more examples, some of which can be traced back to the *Principia:* a body falling in a uniform gravitational field (Galileo's problem), a mass subject to a linear restoring force (Hooke's law), a mass on a spring in a viscous medium (a damped harmonic oscillator), and so on.

Proceeding to motion in two or three dimensions, one meets the problem of a body subject to a central force varying inversely as the square of the distance from the center of force. The solutions are illustrated by the motion of a planet around the sun or a satellite around the earth. Such problems, of course, occupy much of the *Principia.* And they are historically important because they invoke the force of gravity in the form of Newton's "law of universal gravitation." This is certainly the most famous "law" in the whole history of science, and it is generally described as such in modern texts.

The above progression is typical, but many other orders of presentation and groupings of topics can be found. Some texts, for example, devote separate chapters to the conservation of momentum and the conservation of energy. Yet whatever the particular order of presentation or grouping of topics, there is a remarkably common set of core topics and examples. The conclusion that many of these core examples are the historical descendants of Kuhnian exemplars is irresistible.[3]

Beyond the cases mentioned the selection of topics varies from author to author. Most texts include a chapter on the motion of rigid bodies, the most important result noted being that for many purposes a rigid body can be treated as a particle located at the center of mass of the rigid body. This

CONTENTS

	PAGE
PREFACE	v

Chapter I

THE LINEAR MOTION OF A PARTICLE

Introduction	1
1. The Dynamics of Rectilinear Motion	4
2. The Energy Integral	9
3. Qualitative Use of the Energy Integral	11
4. The Conservation of Energy	13
5. Units	16
Problems	17

Chapter II

THE LINEAR OSCILLATOR

Introduction	21
1. Free Oscillations	24
2. The Rate of Damping, and Q	27
3. Forced Oscillations and Resonance	29
4. Superposition of Transient and Forced Oscillation	33
5. Forced Motion under General External Forces	36
6. The Energy and the Linear Oscillator	38
Problems	40

Chapter III

MOTION IN TWO AND THREE DIMENSIONS

Introduction	43
1. The Dynamics of Motion in Two or Three Dimensions	44
2. The Conservation of Energy in Two and Three Dimensions	46
3. Motion of a Particle in a Central Field	52
4. The Inverse Square Law	58
5. The Energy Method for the Inverse Square Law	62
Problems	66

Figure 3.1. Contents page of a typical mechanics text. Reproduced from Slater and Frank (1947).

example illustrates the powerful strategy of reducing a new and apparently complex problem to a simpler problem for which a solution already exists.

The selection of further topics tends to break down between those authors who emphasize additional *physical* applications and those who emphasize additional *mathematical* techniques. Examples of further physical topics include fluid mechanics and the motion of continuous media such as a vibrating string or membrane. Additional mathematical techniques include generalized coordinates, canonical transformations, and Hamilton-Jacobi theory. The choice of additional examples is also a function of the authors' interest in making connections with contemporary subjects, particularly quantum mechanics and relativistic mechanics. Concluding a text on classical mechanics with a chapter on special relativity is now fairly common.

Figure 3.1 reproduces the first page of the table of contents of a typical mechanics text. Readers are invited to inspect the rest of this table of contents, or those in other texts, at their leisure. Before attempting a more theoretical characterization of the structure of classical mechanics, let us investigate one example in greater detail.

The Linear Oscillator

Every mechanics text I have ever seen treats the case of a linear restoring force. Here the force on a particle is proportional to the negative displacement of the particle from its rest position. The second law for this case is

$$(3.2) \qquad F = ma = m \, d^2x/dt^2 = -kx,$$

where k is the constant of proportionality. Solving this equation for both x and v as functions of time is a simple exercise in differential equations, and all texts present the solution in greater or lesser detail, depending on the educational level of the intended audience. The solutions for both position, $x(t)$, and velocity, $v(t)$, as functions of time have the same general harmonic form represented by the function

$$(3.3) \qquad f(t) = A \cos(wt) + B \sin(wt),$$

where $w^2 = k/m$ and both B and C are constants to be determined by the "initial conditions." (The general solution to any second-order differential equation must have two free parameters.) Equation (3.3) defines what is called simple harmonic motion. Only the authors of the more advanced, mathematically oriented texts are content with such a "formal" presentation.

HOOKE'S LAW

The vast majority of texts introduce harmonic motion through the example of Hooke's law, which states that the force exerted by a spring is proportional to the amount it is stretched. The constant, $k,$ is then interpreted as a measure of the stiffness of the spring. Figure 3.2 shows a typical representation of the problem.

With a particular application in mind, text authors have some motivation for considering specific initial conditions. Suppose, for example, that the motion is initiated by moving the mass a distance A to the right and releasing it with no initial velocity. The initial conditions, then, are $x(t = 0) = A$ and $v(t = 0) = 0$. In this case the solutions to equation (3.2) reduce to

(3.4) $\qquad x(t) = A \cos(wt) \qquad v(t) = -Aw \sin(wt)$.

These solutions are represented graphically in figure 3.3.

Most authors are careful to indicate that applying equation (3.2) to the setup pictured in figure 3.2 requires some "simplifying assumptions." One text (Wallace and Fenster 1969, 175) goes so far as to list the following "idealizations": (1) The spring is subject to neither internal nor external frictional forces. (2) The spring is without mass. (3) The force-displacement characteristic of the spring is linear. (4) The mass is subject

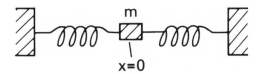

Figure 3.2. The mass and spring system used to illustrate Hooke's law and simple harmonic motion.

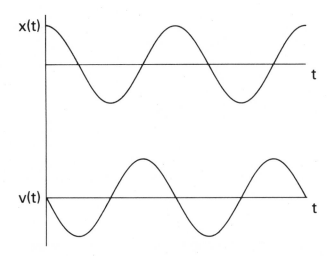

Figure 3.3. Position and velocity as functions of time for the mass and spring system of figure 3.2, with initial conditions $x (t = 0) = A$ and $v (t = 0) = 0$.

to no frictional forces. (5) The wall is rigid, so the wall recoil due to motion of the mass may be neglected.

This list of idealizations is especially interesting because it makes clear that the authors are not talking about any particular real mass-spring system. There are no springs without any mass whatsoever or without internal frictional forces. Instead, the authors are dealing with an ideal mass-spring system that perfectly satisfies equation (3.2). The idealizations are required to ensure that the conditions of the equation are satisfied.

THE SIMPLE PENDULUM

As a further illustration of simple harmonic motion, many texts discuss the simple pendulum. Figure 3.4 depicts a pendulum of length l subject to a uniform gravitational force, $-mg$. A pendulum, of course, moves in at least two dimensions: horizontally and vertically. The problem can be reduced to one dimension, however, by considering only the horizontal, x, component of the motion. The downward gravitational force, $-mg$, is partially balanced by the tension along the string, S, which has magnitude mg $\cos(a)$. S may in turn be resolved into a vertical and horizontal component, the horizontal component being $-S \sin(a)$. Since $\sin(a) = x/l$, the equation of motion in the horizontal direction is

$$(3.5) \qquad m\, d^2x/dt^2 = -mg \cos(a) \sin(a) = -(mg/l)x \cos(a).$$

This, however, is not quite the equation for simple harmonic motion (equation 3.2).

At this point a convenient approximation is introduced. Assume that the angle of swing, a, is small enough so that $\cos(a)$ is approximately equal to

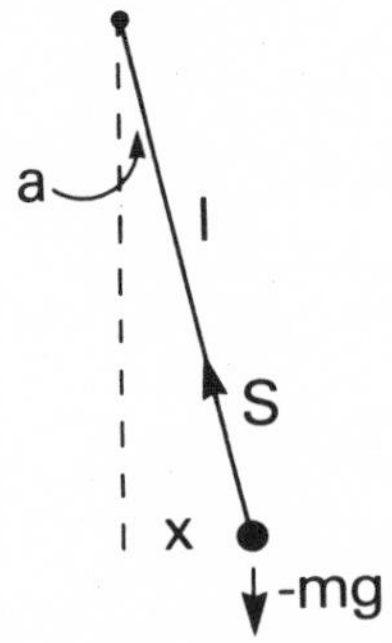

Figure 3.4. A simple pendulum.

one. This assumption is sometimes formally justified by the fact that the function cos (a) is equal to the infinite sum

(3.6) $$\cos(a) = 1 - a^2/2! + a^4/4! - a^6/6! + \ldots .$$

Thus, setting $\cos(a) = 1$ neglects the second and higher powers of a. With this approximation, equation (3.5) reduces to

(3.7) $$m \, d^2x/dt^2 = -(mg/l)x,$$

where the constant mg/l plays the role of the spring constant in the Hooke's law example. The solutions are as before, with $w^2 = g/l$.

From equations (3.4), or figure 3.3, it is evident that the system makes one complete oscillation in a period of time, T, given by the equation

(3.8) $$wT = 2\pi, \text{ or } T = 2\pi \sqrt{l/g}$$

This equation verifies Galileo's and Newton's results that the period of a pendulum is proportional to the square root of its length and independent of its mass.

The move from the mass-on-a-spring example to the simple pendulum seems to me a clear case of what Kuhn called "direct modeling." The two examples are not just special cases of a general relationship. One manages to reduce the pendulum, a two-dimensional system, to the one-dimensional case only by means of a judicious approximation that restricts the pendulum to small angles of swing. In particular, the step from the original application of Newton's laws to the two-dimensional pendulum (equation 3.5) to the one-dimensional version (equation 3.7) is *not* a matter of purely mathematical, or logical, *deduction*. "Approximation" is a valid rule of deduction only in physicists' jokes about mathematics.

THE HAMILTONIAN FORMULATION

In somewhat more advanced texts the harmonic oscillator is also (or perhaps only) treated by focusing on *energy*. The kinetic energy of a particle at time t is defined as

(3.9) $$KE(t) = 1/2m \, v^2 = (1/2m)p^2,$$

where $p = mv$. The potential energy is a function of position, x, and is defined as the integral of the force from some arbitrary initial position to x. For the Hooke's law problem

(3.10) $$PE(x) = 1/2 \, kx^2 = (k/2)x^2.$$

The "Hamiltonian" for the system, H, is the total energy, that is,

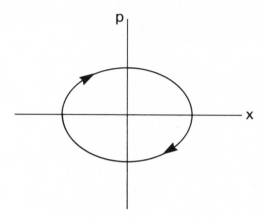

Figure 3.5. The state of a simple harmonic oscillator in position-momentum space.

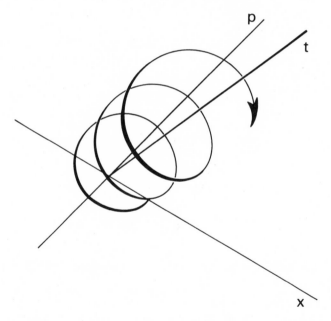

Figure 3.6. The state of a simple harmonic oscillator in position-momentum-time space.

(3.11) $$H = KE + PE = (k/2)x^2 + (1/2m)p^2.$$

The equations of motion for the system can be written in terms of the Hamiltonian as

(3.12) $$dx/dt = DH/Dp, \text{ and } dp/dt = -DH/Dx.$$

Solving these equations yields the standard solutions for the position, $x(t)$, and the momentum, $p(t)$.

The simple harmonic oscillator is a conservative system, that is, its total energy remains constant while oscillating between the potential and kinetic components. If we take x and p as the axes of a two-dimensional, Euclidian state space, the state of the system at any time, which consists of simultaneous values of $x(t)$ and $p(t)$, is represented by a single point. Equation (3.11) implies that the state of the system in this space is confined to an ellipse. Imposing our earlier initial conditions, we find the ellipse has the shape given in figure 3.5. Other initial conditions yield ellipses with other shapes and orientations.

The connection between this state space and our earlier solutions can be seen by imagining a third axis, time, perpendicular to the x-p plane. The state of the system as a function of time is then represented by an elliptically shaped spiral moving out along the t axis, as shown in figure 3.6. Projecting this path on the x-t and the p-t planes yields the sinusoidal functions pictured in figure 3.3. The ellipse in the x-p plane pictured in figure 3.5 is, of course, the projection of the total state in the x-p-t space onto the x-p plane.

THE DAMPED LINEAR OSCILLATOR

To conclude this brief excursion into the world of the mechanics textbook, let us look at just one more case, the damped harmonic oscillator. Real springs and pendulums are obviously not conservative systems since they eventually run down and come to rest. The reason, we say, is that the energy is lost to friction—for example, the internal friction of the spring, air resistance, and friction in the pendulum support. Newton was much concerned with these problems. Book 2 of the *Principia* is devoted to motion in resisting media.

The first case Newton considered assumes that the friction, say, air resistance on the pendulum, is a linear function of the velocity. The second law for a pendulum with friction of this type has the form

(3.13) $$m \, d^2x/dt^2 = -(mg/l)x + bv.$$

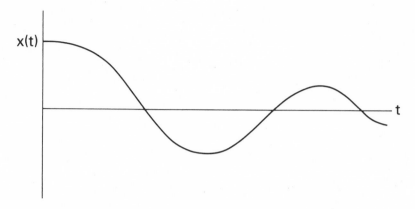

Figure 3.7. The position of a damped harmonic oscillator as a function of time.

The solution for $x(t)$ when b is small (underdamping) is pictured in figure 3.7. Qualitatively, the period is increased from the frictionless case (the pendulum swings more slowly), and the amplitude decreases exponentially in time.

The next complication would be to assume that the pendulum (or spring) is driven by some force that counteracts or even overcomes the frictional force—as in a pendulum clock. The solution, of course, would depend on the exact form assumed for the driving force. If the driving force is itself harmonic, there is the possibility of resonance between the "natural" period of the oscillator and the driver.

Enough! It is time to begin developing a more general, theoretical account of what a theory like classical mechanics might be.

Interpretation and Identification

There were *two* components to the logical empiricists' picture of theories: a purely formal calculus, and "correspondence rules" that link terms in the formal calculus with terms antecedently understood. In the case of mechanics a correspondence rule might say, for example, that x stands for the position of a particle. In one of its more empiricist renditions this account treats correspondence rules as *operational definitions*. Thus, x might be operationally defined as "position measured with a yardstick."

If one's aim is not rational reconstruction and justification, but simply understanding how theories actually function in science, it is relevant that one seldom finds explicit formulations of anything like correspondence rules in standard texts. One does, however, find informal references to the determination of mass, force, position, and momentum. These references might be interpreted as implicit invocations of correspondence rules except for the fact the authors make no distinction between what is "observable" and what is not. The informal references to measuring forces, which philosophers generally regard as "theoretical," are no different in character from those referring to positions, which are supposed to be "observable." The informal remarks, then, are not playing the role the logical empiricists required.

Postpositivist philosophers of science have, of course, been unanimous in rejecting correspondence rules in all their manifestations. Yet any theory of science needs some way of accounting for the phenomenon that motivated their introduction, the fact that scientists do use mathematical symbols to represent things in the real world. Introducing correspondence rules as explicit parts of a theory may be a poor way of accounting for this phenomenon, but one cannot ignore the phenomenon itself.

Actually, there are at least *two* different phenomena we must account for. One is the linking of the mathematical symbols with *general terms,* or concepts, such as "position." A second phenomenon is the linking of a mathematical symbol with some feature of a *specific object,* such as "the position of the moon." I will refer to the former as the problem of *interpretation* and the latter as the problem of *identification.* For example, looking at the formula $F = -kx$, we may *interpret* x as the displacement of a particle from its rest position. In applying the formula to the study of a particular mass on a spring, we *identify* x as the displacement of this particular mass from its equilibrium position.

Kuhn was surely on the right track when he emphasized the importance of *exemplars* in the training of scientists. It is by studying cases like the linear oscillator and discovering what kinds of real systems exhibit harmonic motion that scientists learn how to interpret the mathematical symbolism and how to identify particular instances of that symbolism. This observation, of course, does *not* constitute a *theory* of interpretation or identification. It is not even a theory of how interpretations and identifications are learned. But it is relevant to such a theory in a way that correspondence rules never could be.

The phenomena of interpretation and identification are not unique to science. They occur in any systematic attempt to use language in dealing with objects in the real world. The theory of these phenomena, then, is part of the general theory of language and cognition. In developing a cognitive theory of science, therefore, one will eventually wish to make use of relevant findings from linguistics, and from the cognitive sciences generally, to explain these phenomena. One is unlikely to explain them adequately by purely philosophical theories of meaning or reference.

It would be a mistake, however, to think that one must have a good account of the phenomena of interpretation and identification in hand before one can proceed with the business of constructing a naturalistic theory of science. Scientific activities share much with other human activities, particularly those relying heavily on the use of language. But there are also many things relatively peculiar to science. For the moment I shall focus on these peculiarities.

The Laws of Motion

As already indicated, there is some contemporary controversy over the status of Newton's laws of motion. Newton and his followers in the eighteenth and nineteenth centuries seem to have regarded those laws as general truths about the world. At the end of the nineteenth century it was seriously argued, for example by Mach and Hertz, that one or another of the three laws might better be understood as *definitions*. My interpretation of how classical mechanics is now understood within the physics community will turn out to be somewhat along definitional lines. My immediate concern, however, is to challenge the view that the laws of motion function in classical mechanics as well-confirmed, empirical generalizations.

Interpreting the laws of motion as general, empirical statements played an important role in the logical empiricist view that theories are to be understood as axiomatic systems. The laws of motion, regarded as empirical generalizations, are the prime candidates for the axioms of any such axiomatic system. An examination of even the simplest textbook example shows, however, that, if considered as general, empirical statements, the laws of motion must be judged either as false or at best as irrelevant to the science of mechanics.

THE LAW OF THE PENDULUM

We have already seen that the most general "laws" of mechanics, like $F = ma$, are not really empirical claims, but more like general schemas that need to be filled in with a specific force function. So let us fill in. The classical law of the pendulum (equation 3.8), which is derived from equation 3.7, provides a convenient example. Can this law be understood as a true, universal statement about all pendulums? I think not, and for familiar reasons.

Let us consider a particular pendulum, say, the one in the antique grandfather clock that sits in my living room. Does its motion satisfy the law of motion given in equation 3.7? Obviously not. Equation 3.7 incorporates the approximation that the angle of swing is very small. For any finite angle of swing, equation 3.7 is not strictly true. Well then, what about equation 3.5, which does not make this approximation? Wrong again. That equation assumes a uniform gravitational force. Now, if the earth were a perfectly homogeneous sphere, the force would be uniform. But the continents and oceans make this impossible. Moreover, the heterogeneities of the earth are themselves known to only an exceedingly rough approximation. And then there are more local heterogeneities, such as hills and large buildings. In addition, of course, there is the influence of gravity from other sources, the sun, the moon, and the planets. Thus, even considering just the gravitational forces, equation 3.5 is itself still an approximation. One could think of additions to the equation that might make it a better approximation, but no one would pretend to be able to complete the process so that the equation correctly represented all the existing gravitational forces.

But gravitational forces are not the only forces acting on the pendulum in that clock. There is a frictional force in the joint that attaches the pendulum to the body of the clock. And there is air resistance on the pendulum bob. Textbooks represent frictional forces like air resistance as a linear function of the object's velocity. Yet physicists know that air friction is not a simple, linear function of velocity, but a more complex, nonlinear function. And in the specialized literature one can find more "realistic" functions to represent air resistance. But no one claims any particular function is exactly correct.

I will forgo elaborating on the fact that the pendulum in the clock is not free-swinging but has a driver that gives it a little push on each swing. The general point of belaboring all these details is that if one thinks the laws of motion give a literally true and exact description of even the simplest of physical phenomena, the laws of motion would have to be incredibly more complex than any yet written down—or, given limitations on human knowledge and energy, than could actually be written down. One is therefore reduced to claiming that *in principle* it is (logically?) *possible* to provide true laws of motion. But what role can this claim play if in fact no one ever writes down a true law of motion? The way scientists practice mechanics, it seems that the literal, exact truth of the laws of motion is irrelevant.

The physicist's attitude, well represented in the textbooks, is that the factors mentioned (nonuniformity of gravity near the surface of the earth, the gravitational force of the moon, nonlinearities in air resistance) are all very small compared to the uniform gravitational force of the earth considered as a uniform sphere. Some texts even provide order-of-magnitude comparisons of such forces. But this is just to say that, for the physicist, the literal, exact truth is not what matters. What matters is being close enough for the purpose at hand, whatever that might be.

Those scientists and philosophers who have taken it for granted that the laws of nature are well confirmed, general statements have obviously not been ignorant of the fact that scientists regularly use approximations. But they have taken this to be a relatively inconsequential fact about science. They have regarded the fact as a matter of only practical, not theoretical, importance. This attitude has been reinforced, I think, by taking logical systems and mathematical theories as the model for scientific theories. In logic and mathematics, at least before the age of computers, approximate methods served only the practical end of making calculations possible, or at least easier. Empirical science is different. Idealization and approximation are of its essence. An adequate theory of science must reflect this fact in its most basic concepts.[4]

Models and Hypotheses

Mechanics texts continually refer to such things as "the linear oscillator," "the free motion of a symmetrical rigid body," "the motion of a body subject only to a central gravitational force," and the like. Yet the texts themselves make clear that the paradigm examples of such systems fail to satisfy fully the equations by which they are described. No frictionless pendulum exists, nor does any body subject to no external forces whatsoever. How are we to make sense of this apparent conflict?

MODELS

I propose that we regard the simple harmonic oscillator and the like as *abstract entities* having all and only the properties ascribed to them in the standard texts. The distinguishing feature of the simple harmonic oscillator, for example, is that it satisfies the force law $F = -kx$. The simple harmonic oscillator, then, is a constructed entity. Indeed, one could say that the systems described by the various equations of motion are *socially* constructed entities. They have no reality beyond that given to them by the community of physicists.

It is a useful methodological tactic in constructing a theory of science to employ, insofar as possible, terms that are current in sciences themselves—and with roughly the meaning they have in these sciences. This tactic is useful because all of us who study science share to some extent the cultures of the sciences we study. And we must communicate with the other scientists who practice these sciences. Of course, we shall have to employ some concepts in our science of science that are not found in the cultures of the sciences we study. That is a necessary part of developing an independent science. But we should not create unnecessary barriers to communication with our subjects.

I suggest calling the idealized systems discussed in mechanics texts "theoretical models," or, if the context is clear, simply "models." This suggestion fits well with the way scientists themselves use this (perhaps overused) term. Moreover, this terminology even overlaps nicely with the usage of logicians for whom a model of a set of axioms is an object, or a set of objects, that satisfies the axioms. As a theoretical model, the simple harmonic oscillator, for example, perfectly satisfies its equation of motion.

To avoid confusion, I should note at the outset that we shall have to employ the term 'model' in at least one other distinct sense. Not all theoretical models are models in the further sense of being *exemplars* on which other theoretical models are modeled. Any theoretical model could in principle play this role, but in fact relatively few do, the simple harmonic oscillator being one of the more prominent examples.

The relationship between some (suitably interpreted) equations and their corresponding model may be described as one of characterization, or even definition. We may even appropriately speak here of "truth." The interpreted equations are *true of* the corresponding model. But truth here has no *epistemological* significance. The equations truly describe the model because the model is defined as something that exactly satisfies the equations.

This view of the relationship between linguistic entities, statements or equations, and models, which, even though abstract, are not themselves linguistic entities, is very similar to that advocated by van Fraassen. And it has the same advantages. When viewing the content of a science, we find the models occupying center stage. The particular linguistic resources used to characterize those models are of at most secondary interest. There is no need rationally to reconstruct scientific theories in some preferred language for the purposes of metascientific analysis.

As do the statements used to characterize them, models come in varying degrees of abstraction. At its most abstract the linear oscillator is a system with a linear restoring force, plus any number of other, secondary forces. The simple harmonic oscillator is a linear oscillator with a linear restoring force and no others. The damped oscillator has a linear restoring force plus a damping force. And so on. Similarly, the mass-spring oscillator identifies the restoring force with the stiffness of an idealized spring. In the pendulum oscillator the restoring force is a function of gravity and the length of the string. And so on.

"The linear oscillator," then, may best be thought of not as a single model with different specific versions, but as a *cluster* of models of varying degrees of specificity. Or, to invoke a more biological metaphor, the linear oscillator may be viewed as a family of models, or still better, a family of families of models.

HYPOTHESES

As the ordinary meaning of the word 'model' suggests, theoretical models are intended to be models *of* something, and not merely exemplars to be used in the construction of other theoretical models. I suggest that they function as "representations" in one of the more general senses now current in cognitive psychology. Theoretical models are the means by which scientists represent the world—both to themselves and for others. They are used to represent the diverse systems found in the real world: springs and pendulums, projectiles and planets, violin strings and drum heads.

This understanding of the role of theoretical models in science immediately raises issues of "realism," which have recently much vexed both philosophers and sociologists of science. I shall consider these issues in the next chapter. Here I wish only to attempt a little more precision in describing the relationship between a theoretical model and that of which it is a model. This requires introducing a new concept, that of a "theoretical hypothesis." Here again I intend that the term overlap considerably with the use of the term 'hypothesis' by scientists themselves, although my usage will be more systematic.

Unlike a model, a theoretical hypothesis is, in my account, a *linguistic* entity, namely, a statement asserting some sort of relationship between a model and a designated real system (or class of real systems). A theoretical hypothesis, then, is true or false according to whether the asserted relationship holds or not. The relationship between model and real system, however, cannot be one of truth or falsity since neither is a linguistic entity. It must therefore be something else.[5]

Van Fraassen (1980) suggested that the desired relationship is one of isomorphism. Now, there is surely no logical reason why a real system might not actually be isomorphic to some model. Yet for none of the examples cited in standard mechanics texts, for example, is there any claim of isomorphism. Indeed, the texts often explicitly note respects in which the model fails to be isomorphic to the real system. In other words, the texts often explicitly rule out claims of isomorphism. If we are to do justice to the scientists' own presentations of theory, we must find a weaker interpretation of the relationship between model and real system.

The appropriate relationship, I suggest, is *similarity*. Hypotheses, then, claim a *similarity* between models and real systems. But since anything is similar to anything else in some respects and to some degree, claims of similarity are vacuous without at least an implicit specification of relevant *respects* and *degrees*. The general form of a theoretical hypothesis is thus: Such-and-such identifiable real system is similar to a designated model in indicated respects and degrees. To take a different example:

> The positions and velocities of the earth and moon in the earth-moon system are very close to those of a two-particle Newtonian model with an inverse square central force.

Here the respects are "position" and "velocity," while the degree is claimed to be "very close."

A less stilted formulation of the above hypothesis, one closer to how scientists actually talk, would be:

> The earth and moon form, to a high degree of approximation, a two-particle Newtonian gravitational system.

The latter formulation tends to blur the distinction between the theoretical model and the real system, a distinction that a theory of science should, I think, keep sharp. It also fails to distinguish respects and degrees, lumping both into the vaguer notion of "approximation." But so long as these distinctions are understood, the more relaxed formulation is generally clear enough.

That theoretical hypotheses can be true or false turns out to be of little consequence. To claim a hypothesis is true is to claim no more or less than that an indicated type and degree of similarity exists between a model and a real system. We can therefore forget about truth and focus on the details of the similarity. A "theory of truth" is not a prerequisite for an adequate theory of science.[6]

Although philosophers have often disparaged the notion of similarity as being too vague (Goodman 1970), accumulating evidence from the cognitive sciences, including even the neurosciences (P. S. Churchland 1986), suggests that human cognition and perception operate on the basis of some sort of similarity metric. This reinforces the idea that similarity is a particularly promising relationship for use in a naturalistic theory of science.

Definitions, Models, and Reality

Most theories of science, whether old or new, assume that any representational relationship between theory and reality would have to be understood as a "correspondence" between scientific statements and the world. The fate of any understanding of theories as somehow representing reality has thus been linked to the fortunes of a correspondence theory of truth. It is here that the battle is usually joined. The interpretation I have offered above undercuts these arguments by denying the common assumption. There is, on this account, no direct relationship between sets of statements and the real world. The relationship is indirect through the intermediary of a theoretical model, as pictured in figure 3.8.

Of course, assertions about the existence of similarity relationships between models and real systems do require theoretical hypotheses, which are indeed linguistic entities. But for these a "redundancy theory" of truth is all that is required (see note 6). And there is a truth relationship of the correspondence type between the statements characterizing an abstract model and the model itself. But this correspondence, which reduces to definition, is not a problem. The relationship that does the heavy representational work is not one of truth between a linguistic entity and a real object, but of similarity between two objects, one abstract and one real. From this point of view the difficulties with the standard view arise because it tries to forge a direct semantic link between the statements characterizing the model and the world—thus eliminating the role of models altogether.

What Is a Scientific Theory?

Even just a brief examination of classical mechanics as presented in modern textbooks provides a basis for some substantial conclusions about the overall structure of this scientific theory as it is actually understood by the bulk of the scientific community.

What one finds in standard textbooks may be described as a cluster (of clusters of clusters) of models, or, perhaps even better, as a population of models consisting of related families of models. The various families are constructed by combining Newton's laws of motion, particularly the second law, with various force functions—linear functions, inverse square functions, and so on. The models thus defined are then multiplied by adding other force functions to the definition. These define still further families of models. And so on. Figure 3.9 offers a partial picture of the resulting overall structure of the families of models.

It must be emphasized that figure 3.9 shows only part of the gross structure of the theory. Many features are not pictured. The figure gives no indication, for example, that some models are more specific versions of other, more general models, as when a generalized model with a linear restoring force is reduced to a model of an idealized mass-spring system. It also does not show that these models are not merely mathematical structures, but "interpreted" models. More significantly, figure 3.9 shows only the models and not any of the real systems they may be used to represent. *Hypotheses,* in my technical sense, do not appear in the picture. Nor do the linguistic resources used to define the models.

The coordinate notions of a theoretical model and a theoretical hypothesis provide the main ingredients for a *theory of theories* (Bromberger 1963). So far, however, I have not explicitly said what a theory is. But this, I think, is no longer a significant question. We already have enough con-

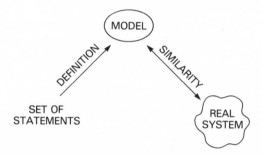

Figure 3.8. Relationships among sets of statements, models, and real systems.

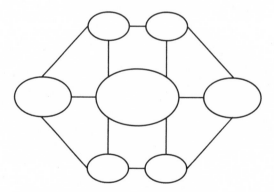

Figure 3.9. A partial representation of the families of models associated with classical mechanics.

ceptual machinery to say anything about theories that needs saying. All we lack is something to which we can attach the (often honorific) label 'theory'. In keeping with the methodological tactic of staying as close to scientific usage as possible, the problem is simply to capture as much of scientific talk about theories as we can using the machinery at hand. But this turns out to be no trivial task because scientific usage pulls in several different directions.

Since it seems that theories are things one finds written down in textbooks, among other places, one is inclined to look for something "linguistic" to be the theory itself. But one cannot identify the theory of mechanics with any particular set of *sentences*. That would make the translation of a textbook from English into French a different theory, which seems absurd. Philosophers long ago introduced the more abstract entities called "statements," or "propositions," to handle just this problem. A proposition is supposed to be, roughly, what a sentence asserts. One can therefore assert the same proposition using different sentences, some in English and some in French, for example.

If one looks for statements, the obvious candidates are Newton's laws of motion and the various force laws one finds in the standard texts. But if they are understood as statements making claims directly about the world, all the laws of motion and force laws one finds written down are known to be false—a discomforting fact to say the least. That is largely why I have argued that these laws are to be interpreted as providing definitions of various models, models that are nonlinguistic, though abstract, entities.

On the other hand, if understood as definitions, the laws of motion make no claims about the world. Now physicists, as we have seen, are ambiguous about whether Newton's laws are empirical claims or definitions. But it would be a rare physicist who would agree that the whole theory consists of nothing but definitions. Newtonian mechanics, most physicists would insist, does make claims about the world, and anyone who tries to say otherwise will get short shrift.

An obvious alternative is to focus on theoretical *hypotheses*. These are appropriately linguistic, and they make claims about the world—they may be true or false. The trouble with this suggestion is that Newton's laws and the force laws turn out not to be among the statements that are part of the theory of Newtonian mechanics. Consider the most outlandishly general hypothesis possible: The whole universe is similar to a Newtonian model defined as follows. . . . This hypothesis incorporates Newton's laws, but the laws themselves fail to appear as separate statements making up the theory.

A compromise is to say that a theory includes both statements defining the population of models and hypotheses claiming a good fit between various of the models and some important types of real systems. The price we pay for trying to have our cake and eat it too is that a theory turns out to be a rather heterogeneous type of thing. It includes both definitions and empirical hypotheses. But that may be a small price for capturing the diverse intuitions of what a theory is.

My only objection to this compromise is that it puts too much emphasis on matters linguistic. It focuses attention on the statements that define the population of models rather than on the models themselves. I would prefer to substitute the models for the definitions. Newton's laws and the force laws would remain, though only implicitly, and not in linguistic garb. They would be embodied in the models.

My preferred suggestion, then, is that we understand a theory as comprising two elements: (1) a population of models, and (2) various hypotheses linking those models with systems in the real world. Thus, what one finds in the textbooks is not literally the theory itself, but statements defining the models that are part of the theory. One also finds formulations of some of the hypotheses that are also part of the theory. That characterization seems to me sufficiently close to how physicists think and talk to be useful. If, however, someone prefers the earlier compromise, or some near variant, I would not strongly object. The difference is of little substance.

Figure 3.10 is an attempt to picture the structure of a theory such as classical mechanics. It is helpful to compare this picture with the logical empiricist picture shown in figure 2.1. The ellipses in figure 3.10 represent not "terms" or "concepts," but whole models—abstract, nonlinguistic en-

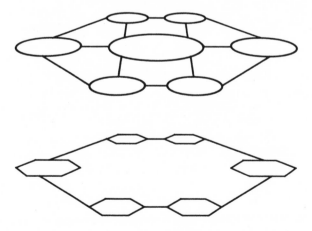

Figure 3.10. A picture of a theory, showing a family of models and related real systems.

tities. The links between models are not logical connections, but relations of similarity. In some cases the difference between two models is that one is an approximation of the other—again not a logical relationship. The links between models and the real world below are nothing like correspondance rules linking terms with things or terms with other terms. Rather, they are again relations of similarity between a whole model and some real system. A real system is *identified* as being similar to one of the models. The *interpretation* of terms used to define the models does not appear in the picture; neither do the defining linguistic entities, such as equations.[7]

ARE THEORIES WELL-DEFINED ENTITIES?

It is a consequence of the above interpretation that a scientific theory turns out not to be a well-defined entity. That is, no necessary and sufficient conditions determine which models or which hypotheses are part of the theory.

This is most obvious in the case of hypotheses. I suspect that most scientists would agree that general hypotheses asserting similarities between planets and Newtonian models with inverse square central forces are part of the theory. Many would also agree if one of the planets was designated as the earth. But what about the claim that the pendulum on my antique clock resembles a model of a harmonically driven linear oscillator? Is that claim part of the "theory" of classical mechanics? Here it seems that one could argue the case either way. My view is that it matters not at all how, or even whether, one answers such questions.

The population of models for classical mechanics is also not well defined because there are no necessary and sufficient conditions for what constitutes an admissible force function. This raises an interesting question. What determines whether a model is to count as a proper Newtonian model?

One answer is that to be part of the theory of classical mechanics a model must bear a "family resemblance" to some family of models already in the theory. That such family resemblances among models exist is undeniable. On the other hand, nothing in the structure of any models themselves could determine that the resemblance is sufficient for membership in the family. That question, it seems, is solely a matter to be decided by the judgments of members of the scientific community at the time. This is not to say that there is an objective resemblance to be judged correctly or not. It is to say that the collective judgments of scientists *determine* whether the resemblance is sufficient. This is one respect in which theories are not only constructed, but socially constructed as well.[8]

What about Axiomatic Presentations of Mechanics?

The structure of mechanics found in standard, professional-level textbooks is almost surely *not* that of a linear arrangement of theorems derived from a given set of axioms—with applications being "instances" of the various theorems. That is simply not the way the theory is typically presented. But even the logical empiricists did not claim that scientists actually think of theories axiomatically. Their view was that, in some sense, theories "really are" interpreted axiomatic systems, whether scientists typically conceive of them that way or not. And in fact one can construct axiomatic systems that capture at least some aspects of classical mechanics.[9]

A typical way to give classical mechanics an axiomatic structure is to narrow one's focus to a particular force function such as Newton's law of universal gravitation. Putting this force law together with Newton's laws of motion yields a workable set of axioms from which one can derive some interesting theorems. Yet this case pulls in far less than one tenth of the contents of a typical text. To capture more of the theory in this manner would require a dozen or more similar axiomatic systems, each incorporating a different force function.

A more inclusive axiomatic approach is suggested by high-level mathematical treatments such as that of Abraham and Marsden (1978). One can begin with a formulation of mechanics using generalized force functions that are, abstractly, functions of several variables such as position, velocity, and time. One then deduces various results using only this very generalized characterization of the force function. At some point one generates numerous special cases by introducing various restrictions on the particular form of the force function, for example, that it be a function of position only. The result is an axiomatic development with many branches, each a different special case.

The possibility of constructing a branching axiomatic structure suggests a different interpretation of the axiomatic exercise than that usually found. Rather than regarding the axioms and theorems as empirical claims, treat them all merely as definitions. Then the whole axiomatic structure becomes primarily a systematic way of generating characterizations of families of models. In constructing figure 3.10, I deliberately suppressed the linguistic resources used to characterize the models, leaving behind only their embodiment in the models. One could regain the linguistic component by adding a three-dimensional branching structure above the models pictured in figure 3.10. One need only match up the tips of the branches with the corresponding models.

Although the resulting, more complicated picture of a theory is fairly appropriate for classical mechanics, I would not employ it as a picture for theories in general. Not all things we should want to call theories have a structure so tightly knit that most of the families of models can be generated in a simple, axiomatic way. Nor does an axiomatic presentation of the models make the theory any more well defined in general. It only gives the appearance of doing so. Decisions as to what kinds of models belong in the population are built into the form of the original generalized force function. Further decisions are made later when restrictions on the variables in the force function are introduced to generate special cases. The axiomatic presentation does not eliminate these decisions; it only disguises them.

I would not deny that an axiomatic presentation of the models might provide useful insights into questions about the science of mechanics. But I strongly deny that such an account can in any way be taken as capturing the "real," or "correct," structure of classical mechanics. And I even more vehemently deny that it provides a correct picture of how the theory is in fact understood, and used, by most scientists.

OTHER EVIDENCE

My main evidence for claiming that an axiomatic organization of the models of classical mechanics plays little role in the actual understanding of the theory by most scientists has been the fact that standard professional level textbooks are not organized axiomatically. And it is from those texts that most scientists gain their understanding of the theory. Here I would like to cite briefly two examples of other evidence for the same conclusion, both from the literature on human problem solving.

One example comes from studies of expert chess players (Newell and Simon 1972). One can think of the rules of chess as a set of axioms that can be used to generate all possible games. If chess players thought about chess axiomatically, one would expect them to analyze a game setup with reference to the rules. That is not at all what one finds, however. Rather, the chess master compares the game presented with a remembered game pattern and then proceeds to analyze the current problem in terms of that remembered pattern. Standard estimates of the number of game patterns that a grand master can recall are around 50,000. I would guess that the number of models available to the expert in classical mechanics must be one or two orders of magnitude smaller.

The second example appears in the few studies of how physicists actually solve typical problems in classical mechanics. One of the main devices employed in those studies is to have the scientist think aloud while solving the problem. The researcher then analyzes the protocols generated by the scientist. From a few published reports it seems that physicists do not begin by writing down Newton's equations, adding a force function, and deducing. Rather, they first select from memory a representation of the problem—a model—and work from there. What is retrieved from the physicists' long-term memory, it seems, is not the axioms of mechanics, but an appropriate model. In a summary of recent research focusing on the difference between expert and novice performance in solving physics problems, Larkin et al. (1980, 1342) wrote:

> In every domain that has been explored, considerable knowledge has been found to be an essential prerequisite to expert skill. The expert is not merely an unindexed compendium of facts, however. Instead, large numbers of patterns serve as an index to guide the expert in a fraction of a second to relevant parts of the knowledge store. This knowledge includes sets of rich schemata that can guide a problem solver's interpretation and solution and add crucial pieces of information. This capacity to use pattern-indexed schemata is probably a large part of what we call physical intuition.

This evidence, however, might not be regarded as sufficiently independent of my earlier examination of the structure of standard textbooks. The physicists who served as subjects in this line of research no doubt learned their mechanics from similar texts. It therefore would not be surprising to find that their knowledge of mechanics is organized along the same lines as the textbooks. Nevertheless, it is a tempting speculation that the physicists studied fit this account not simply because the account mirrors the way mechanics textbooks are organized. Rather, I would argue that the modern physics textbook evolved to its present form because this form is well adapted to the actual operations of human cognitive capacities.

Beyond Classical Mechanics

So far my interpretation of what a scientific theory amounts to rests on a single case—classical mechanics. At the beginning of this chapter I gave my reasons for thinking this an exemplary case. By now I would hope that readers familiar with other sciences could readily imagine for themselves how they would go about drawing similar conclusions about their own favorite theories.[10]

The main general lesson is: When approaching a theory, look first for the models and then for the hypotheses employing those models. Don't look for general principles, axioms, or the like. Of course, pay close attention to what scientists write and say. Models are typically characterized using words, though diagrams are often equally important. But language is merely a means to an end—the characterization of a population of models.

In later chapters I will myself apply this lesson to some aspects of nuclear physics and to geology. I will conclude this chapter by citing a recent study of quantum theory.

One might agree with my account of how contemporary scientists understand classical mechanics but disagree that it provides an adequate model for contemporary physics, let alone contemporary science in general. My sources are textbooks written for the most part since World War II. These appeared at least a generation after physicists had agreed that classical mechanics had been superseded by relativity and quantum mechanics. Perhaps the contemporary treatment of classical mechanics is peculiar because it is recognized not to provide the most accurate account of any phenomenon. This failing might even explain the reluctance of contemporary authors to place great emphasis on traditional universal generalizations. These are now known to be false.

Yet even more radical conclusions than mine have recently been based on an examination of contemporary quantum mechanics. In *How the Laws of Physics Lie* (1983), Nancy Cartwright argued that the fundamental laws of modern physics, such as the Schroedinger equation, simply are not true. They lie. To support her view, she exhibited cases in which the Schroedinger equation has been used to develop incompatible accounts of a single phenomenon, such as radiative damping. A unitary theoretical treatment for many important phenomena, she claimed, simply does not exist. But this is just what one would expect if contemporary quantum mechanics is anything like my interpretation of contemporary treatments of classical mechanics.

I disagree only with Cartwright's general description of her conclusion, not its real content. In my view the general laws of physics, such as Newton's laws of motion and the Schroedinger equation, cannot tell lies about the world because they are not really statements about the world. They are, as Cartwright herself sometimes suggests, part of the characterization of theoretical models, which in turn may represent various real systems. But only part of the characterization. There is no real system for which the basic form of the Schroedinger equation by itself describes a model, no more than $F = ma$, by itself, defines a model of anything. One always needs more details, specific force functions, approximations, boundary conditions, and so on. Only then does one have a model that can be compared with a real system.

Cartwright's treatment supports my prescription for how to approach a scientific theory. It should not be difficult to identify the main exemplars of quantum mechanics, such as a single particle in an infinite potential well, around which families of models have been constructed. Quantum mechanics, like classical mechanics, can easily be seen, I am sure, not so much as a unitary formal system, but as a family of families of models. Whether the Schroedinger equation provides as tight a bond among the families as Newton's second law does for classical mechanics is unclear. But it does not matter. The possibility of generating all the families of models from a simple set of general axioms is not a necessary condition for any scientific theory.

4

Constructive Realism

Everyone who has puzzled over the nature of scientific theories has struggled to understand how theories relate to the world. This statement includes both philosophers and sociologists. In this chapter the emphasis is on philosophical concerns. In the next chapter the emphasis shifts somewhat more toward matters sociological.

Most philosophical objections to realism can be put into one of two general categories: conceptual and epistemological. The *conceptual* objections question whether realism can be formulated as a coherent thesis that is neither vacuous nor obviously false. Many of those objections could also be called semantic because they focus on the meaning of 'truth' for theoretical claims. In particular, does a correspondence theory of truth make sense for theoretical claims?

The *epistemological* objections, by contrast, typically grant that realist claims make sense. They do question, however, whether there could, in general, be adequate justification for realist claims. In particular, can one justify any inference from experimental success to the truth of a theoretical hypothesis?[1]

In this chapter I shall be concerned mainly with the conceptual sorts of objections. My aim will be to formulate a version of realism that is conceptually coherent and neither trivially true nor obviously false. The same process also promotes the formulation of various alternative views falling in the anti-realist camp. In the final section of the chapter I will offer replies to several specific conceptual objections. The epistemological sorts of objections will be faced in the next chapter and in subsequent chapters.

Respects of Similarity

Scientists construct theoretical models that they intend to be at least partial representations of systems in the real world. As discussed in the previous chapter, the primary relationship between models and the world is not truth, or correspondence, or even isomorphism, but *similarity*. A theoretical hypothesis asserts the existence of a similarity between a specified theoretical model and a designated real system. But since anything is similar to anything else in some way or other, the claim of similarity must be limited (as least implicitly) to a specified set of respects and degrees.

In recent years many philosophical differences between realist and anti-realist interpretations of science have been formulated in semantical terms like 'truth' and 'reference'. In what sense, if any, can a scientific hypothesis be said to be "true"? Do theoretical terms genuinely "refer"? My interpretation bypasses these semantical questions and focuses directly on the respects and degrees of claimed similarity between model and real system.

The respects in which similarity may be claimed can only be those represented in the model. One cannot claim, for example, that a mechanical system is similar to a classical model with respect to color simply because there is nothing which represents color in any classical model. The models themselves provide an upper limit on the respects in which similarity can be claimed.

Not so with degrees of similarity. The models themselves put no restrictions on the degree to which similarity might be claimed. One might claim, for example, that the successive positions of a pendulum bob are *exactly* as represented in a specified model. Any asserted degree of similarity less than perfection must, therefore, be determined by something other than the model itself. I will consider what this additional something might be a little later. For the moment I shall focus on *respects* of similarity. It is respects of similarity, not degrees, that primarily separate realists from anti-realists.

Although models themselves provide an upper limit on claimed respects of similarity, one may consistently limit claims of similarity between a model and reality to as few aspects of the model as one desires. In general, realists are relatively liberal in the range of respects in which they will attribute similarity between a model and a real system. Empiricists are rather more restrictive. Some constructivist sociologists seem prepared to deny claims of similarity in any respects whatsoever. Models, in this view, are not representations at all.

Mention of constructivism raises a question about my use of the term 'constructive realism'. To many ears it may sound like a contradiction. In fact it originated as a realistic alternative to van Fraassen's "constructive empiricism" (Giere 1985b). The term emphasizes the fact that models are deliberately created, "socially constructed" if one wishes, by scientists. Nature does not reveal to us directly how best to represent her. I see no reason why realists should not also enjoy this insight.

One must remember, however, that constructive realism is a doctrine only about the nature of scientific models and hypotheses, that is, only about scientific representations. It is not a doctrine about scientific judgment, that is, about how scientists judge which models best represent the world. Constructive realism is compatible with these judgments being made in accord with a priori rules of rational choice or by means of purely social negotiations. My claim will be that scientific judgment is a natural, cognitive process. The resulting view is a naturalistic, constructive realism.

Varieties of Empiricism

The general strategy of all forms of empiricism is to restrict claims of similarity to just those respects in which our models correspond to empirical aspects of the world. In van Fraassen's terms hypotheses are restricted to claims about the "empirical substructure" of models. That is all "empirical adequacy" requires. The rest, the theoretical superstructure, may be in some sense "accepted," but not genuinely believed to correspond to anything in the world.

What distinguishes the empirical from the nonempirical? The almost universal answer is observability. Thus, for example, most present-day empiricists, such as van Fraassen, would allow that the positions of a normal pendulum bob are observable. They would therefore be willing to assert that the real pendulum is similar to a theoretical model with respect to claims about successive positions of the bob. But they would not assert similarity for other aspects of the model, for example, the uniform downward gravitational force, which in the model has magnitude $-mg$. Forces have always been a prime candidate for something that is not observable.

Given a distinction between what is observable and what is not, we can distinguish several varieties of empiricism: (1) empiricism that limits claims of similarity to just those aspects of real systems that have in fact already been observed; (2) empiricism that limits claims of similarity to all aspects of real systems that have been, are now, and ever will be observed; (3) empiricism that limits claims of similarity to those aspects of real systems that are *observable,* whether or not they ever are in fact observed. One can find plausible candidates for each of these positions among past and present philosophers. Van Fraassen, for example, favors the more liberal, third variety of empiricism based on observables.[2]

Is the Observable a Useful Category?

One of the major philosophical objections to empiricism has been the viability of the required distinction between observational and nonobservational aspects of the world. The distinction was in fact much more secure when the word 'observational' was identified with what could be subjectively experienced. Sense data at least provided a homogeneous class of things that are uniquely observable. Hardly anyone today, however, is willing to defend so radical an empiricism. This forces the modern empiricist to draw the limits on what is observable somewhere between pendulum bobs and electrons.

The logical empiricists, of course, attempted to draw the distinction linguistically. Some terms, they said, are "observation terms." Thus was born "the problem of theoretical terms." This strategy has many undesirable consequences, and modern empiricists like van Fraassen therefore avoid it. 'Black', for example, turns out to be an observation term applied to an eight ball but not to a mite (Suppe 1974).

Van Fraassen suggested we take it as an empirical problem to determine what average humans can in fact observe with their unaided senses. But even this ploy has strange consequences. Imagine a small but clearly visible mass suspended in equilibrium position between two relatively stiff springs. Once set in motion, however, the mass becomes a complete blur. No one with normal vision can observe when it passes through its original equilibrium position or, indeed, which way it is moving at any instant. Are we to say the position of the mass is observable when it is at rest, but not when in motion?

More important than these philosophical arguments is the fact that the required distinction is simply not found in the practice of science. Textbooks on mechanics, for example, do not distinguish the observability of forces as opposed to positions. Nor are any difficulties of principle raised about measuring forces as opposed to positions. Occasionally, one finds discussion of the typically philosophical issue whether forces can be defined in terms of other quantities, like mass and acceleration. But this is an entirely different issue from the question of observability.

Scientists do, of course, raise questions about what *is* observable. Astrophysicists, for example, have discussed the fact that high-energy solar neutrinos can now be observed, but lower energy neutrinos, which, according to standard models, make up the bulk of the solar neutrino flux, are not observable with existing instruments. This is obviously not the kind of distinction empiricist philosophers have had in mind. In astrophysics 'observable' means something like "reliably detectable with existing instrumentation." By any empiricist standards high-energy neutrinos are as unobservable as their low-energy cousins.[3]

I shall not pursue such arguments further here. My objective at the moment is not so much to refute empiricism as to make sense of realism.

Unrestricted Realism

Having outlined various versions of empiricism, I shall now survey several possible versions of realism. In the framework developed so far, the most straightforward realistic interpretation of any theoretical hypothesis would be as an assertion of similarity between the real system and *every* aspect of the model. Let us call this view "unrestricted realism." It is easy to see, however, that unrestricted realism is too strong a view.

Returning to mechanics, consider a representation of an apple falling from a tree by a simple model of free fall in a uniform gravitational field. Let the height from the ground be represented by the variable y. The relationship between the height, h, from which the apple falls with zero initial velocity and the time taken to reach the ground is given by the familiar formula $h = 1/2 \ gt^2$. If h is given, we can solve for the time to reach "ground zero," which is $t = \pm\sqrt{2h/g}$.

We of course disregard the solution in which time has a negative value. If we regard negative values of time as aspects of the model, we are not practicing unrestricted realism. We restrict claims about the apple to positive values of time. But it may reasonably be claimed that the negative solutions are not to be regarded as "in the model" in the first place. After all, we have an *interpreted* model, and t is interpreted as "time," which, we normally say, has only positive values.

The same, however, cannot be said for positive values of t *greater than* $\sqrt{2h/g}$. These values correspond to negative values of y, the position of the apple. In general, there is nothing wrong with negative values of position in models of free fall. They are merely an artifact of where one chooses to put the origin of the coordinate system. Thus, the general interpretation of the variable y as position does not require that it take only positive values. But in this particular application the origin corresponds to the place where the apple hits the ground. It cannot go any further. Time may go on, but the apple cannot. We use this extra knowledge to limit application of the model in this particular problem.

Anyone familiar with other branches of physics can think of many similar examples in which features of the model are not thought to have any counterpart in the world. I doubt there is any general rule about this. And I am almost certain there is no purely formal rule. That is, nothing in the formal, mathematical structure of the models distinguishes between features we will assume to have a real counterpart and those we will not. And even if we add standard "interpretations" of the variables, often we still have to ignore some features of any model for specific applications.

On the other hand, sometimes taking seriously formal aspects of a model ordinarily thought to have no real counterparts can lead to important discoveries. This was apparently the case with Dirac's discovery of the positron, which appeared first as the representation of an electron with a negated negative charge.

The view I call constructive realism, then, is intended to be a *restricted* form of realism in the sense that theoretical hypotheses are interpreted as asserting a similarity between a real system and some, but not necessarily all, aspects of a model. The question of which aspects, and why not others, is left to be resolved on a case-by-case basis by scientists themselves.

THE CHARYBDIS OF REALISM

Since its inception quantum theory has provided a focus for discussions of realism by both philosophers and physicists. In general, its influence has been at least to soften realist pretensions if not to foster various forms of anti-realism, including positivism. This tradition of employing quantum theory in the service of empiricism is exemplified in a recent article by van Fraassen (1982) bearing the above title.

The target of van Fraassen's attack is the principle of common cause espoused by many realists such as Salmon (1975). Briefly, the principle is this: If two types of events are correlated, either one is the cause of the other, or they are both the effect of a common cause. Van Fraassen argues quite convincingly that the recent experimental verification of Bell's inequality shows the principle of a common cause is violated in some quantum processes. As van Fraassen (1982, 35) put it: "There are well-attested phenomena which cannot be embedded in any common-cause model." The realist is thus left with the choice between rejecting a well-attested physical theory and rejecting the principle of common cause.

As I wish to understand it, constructive realism incorporates no general principles, like that of common cause, that any adequate scientific theory must follow. It is not the job of a theory of science to legislate, a priori, the form scientific theories must take. Nor, I would add, can anyone outside of science hold such authority. I am therefore quite willing to agree with the judgment of physicists that there seem to be some correlations without any physically possible common cause. The most one can say is that this judgment is unprecedented in the history of science. The search for common causes has very often been strikingly successful. But if good evidence shows that quantum processes are an exception, so be it. That may be the way the world is.

The thrust of constructive realism is in the other direction. It rejects all attempts to place general, a priori restrictions on which aspects of scientific models may or may not be asserted to resemble features of real systems—the empiricists' restriction to observable aspects of reality being a prime example.

Metaphysical Realism

Hilary Putnam has accused scientific realists of holding to a "metaphysical realism," which asserts that "there is exactly one true and complete description of 'the way the world is'" (1981, 49). His rejection of realism consists of arguing that metaphysical realism is incoherent. Putnam's formulation of metaphysical realism is in a linguistic rather than a model-theoretic framework. Let us see what sense we can make of such a view in my suggested framework.

Note, first of all, that metaphysical realism is not a thesis about any theory now known. It is about some possible theory. Unlike classical mechanics, this possible theory could not consist of a family of related models. It would have to be just one big model. Every aspect of this model would have to correspond to a feature of the world, and there could be no feature of the world left out. Moreover, the similarity relationship would have to collapse into perfect isomorphism. Finally, the model would have to be unique. No other model could do an equally good job. Thus, metaphysical realism is not only "unrestricted," but also "complete," "perfect," and "unique."

As I understand it, Putnam's main argument against metaphysical realism is that the uniqueness requirement is impossible to satisfy. And the impossibility must be a logical, or conceptual, impossibility because Putnam's argument is purely logical, or conceptual. Now, one can question whether Putnam's argument is valid, but in the present context the point is moot.

Perhaps Laplace was a metaphysical realist. Perhaps some recent philosophers, including Putnam himself a decade ago, have been metaphysical realists. But metaphysical realism plays no role in modern science. On my analysis a major exemplar of a scientific theory, classical mechanics, is not even regarded as being fully realistic, let alone complete, perfect, and unique. And then there is quantum theory. The rejection of metaphysical realism therefore eliminates nothing that an adequate theory of science might require.

Modal Realism

Most recent debates between empiricists and realists have focused on the distinction between empirical and theoretical aspects of reality. It seems to me, however, that *modal* claims, claims of possibility and necessity, provide an even more critical dividing line. Van Fraassen provided a good index of the relative importance of these two issues. He was willing to grant that real systems may, in fact, possess the theoretical structure of scientific models. He merely insisted that no one could ever justifiably assert that resemblance. Regarding modality, however, he was not merely agnostic but atheistic. "The locus of possibility," he insisted (1980, 202), "is the model, not a reality behind the phenomena." In short, possibilities and necessities are only figments of our models—useful, perhaps, but not even candidates for reality.

Modality is important because of its close connection with *causality*. One way of understanding at least some aspects of causality is to regard the modal structure of scientific models as representing a causal structure in real systems. Modal realism is the view that, in some cases at least, a causal counterpart of the modal structure of a scientific model may exist in nature.

Causality in Mechanical Systems

Mechanical systems have long been exemplars of causal systems—at least in Western culture. Let us therefore focus once again on the Hooke's law model of a simple harmonic oscillator, this time in its Hamiltonian formulation. While the Hamiltonian version of Newton's laws in fact fits van Fraassen's own preferred form for presenting theories, it nevertheless also provides an excellent framework for expressing the claims of modal realism. To eliminate controversy over theoretical aspects of the system, let us concentrate on position as the variable of interest.

Everyone agrees that the *model* exhibits a modal structure. For example, the model allows a wide range of different *possible* initial conditions, only one of which can be exhibited at any given time. For each of these (continuously many) different possible sets of initial conditions, one can calculate, using the equations defining the model, a trajectory for the system in the state space, as pictured, for example, in figure 3.6.

Now let us shift our focus from the model to a real mass-spring system. Suppose that the real mass had in fact been started in motion with zero initial velocity. Van Fraassen's constructive empiricist and my constructive realist would agree that this real mass should resemble the model with respect to its actual positions in time. But what about the possible positions that would result from different initial conditions, such as some nonzero initial velocity. Does our claim of similarity between model and real system extend to the modal aspects of the model? Do we claim that the real system possesses these possibilities as well? Or, to use a more traditional philosophical formulation, is it true of the real system, as well as of the model, that its positions would follow some definite other pattern if its initial conditions were different?

Modal realism embraces the affirmative answer to these questions. Modal anti-realists, like van Fraassen, could also be called *actualists*. They claim that science aims only at discovering similarities between our models and the actual histories of real systems. The other possibilities are not inherent in real systems, but exist merely in our models.

Empiricist Objections

Empiricist arguments against modal realism are apparently stronger than those against theoretical realism. There is no possibility that one could observe states of a real system that might have been, but were not in fact, realized. It is not as if someone might invent a new type of microscope that would expose these possibilities to direct observation.

But this argument is conclusive only if one accepts the empiricist doctrine that ultimately all evidence rests on direct observation. If we allow that there might be other bases for making modal claims, the argument ceases to be conclusive. The satisficing decision strategies described in chapter 6, for example, permit the acceptance of modal hypotheses.

Another argument against modal realism, as Quine (1953) has long delighted in pointing out, is that it is often difficult to individuate possibilities. Often, yes, but not always. Many models in which the laws are expressed as differential equations provide an unambiguous criterion for individuating the possible histories of the model. These histories are the trajectories in state-space corresponding to all possible initial conditions. Threatened ambiguities in the set of possible initial conditions can be eliminated by explicitly restricting this set in the definition of the theoretical model. Nor is this criterion limited to models defined in terms of differential equations. A clear distinction between system laws and parameters or initial conditions is generally sufficient. Of course, even classical physics presents cases in which the specification of boundary conditions is ambiguous. But the ambiguity is not nearly so great as a Quine, or a van Fraassen, would suggest.

One could, of course, defend an actualist version of constructive realism. This would be the view that scientific models represent only the actual behavior of both empirical and theoretical aspects of real systems. I will, however, adopt a modal realist interpretation. In spite of philosophical arguments to the contrary, this seems to me the empirically best supported account of theoretical hypotheses.

MODALITY AND CAUSALITY

The oscillating spring is a causal system if anything is. The theoretical model shows us that the frequency of oscillation, f, is functionally related to the ratio, k/m, and functionally independent of the amplitude, A. If the functional relationship obtains merely between the *actual* values of these quantities, it is difficult to see what more there could be to *causality* than merely this functional relationship—as Russell (1912–13) long ago pointed out.

Empiricists have traditionally sought to ground the causal claim in universal generalizations, for example, the claim that *all* oscillating springs exhibit these relationships. On examination such generalizations turn out to be either false or vacuous. For the modal realist the *modal* structure of the model represents, to some degree of approximation, the *causal* structure of the real system. For any real system of the relevant type, then, the functional relationships among the actual values of f, m, k, and A represent causal relationships not because they hold among the actual values in all such real systems, but because they hold among all the possible values in this particular system.

NECESSITY AND PROPENSITY

I have introduced the modal version of constructive realism in the context of deterministic systems of the kind described by classical mechanics. This understanding of how theoretical hypotheses are used is, however, in no way tied to classical mechanics or even to deterministic theories. If one wishes to allow the possibility of genuinely stochastic systems, as quantum systems are widely believed to be, there is a natural extension of strict causal necessity that generally goes by the name "propensity."[4]

Probability models are used in nearly all the sciences, ranging from quantum physics through anthropology. No doubt their wide range of applicability is partly due to their exceedingly simple structure. But what do probability models represent?

The simplest empiricist answer is that they represent actual relative frequencies of kinds of individuals in finite populations. This is the probabilistic analogue of saying that strict causal necessity consists of universal occurrence in all cases. In fact, because of the mistaken (I think) belief that there should be a strict isomorphism between a model and what it represents, most empiricist philosophers postulate hypothetical limiting relative frequencies in infinite sequences to be the real world counterparts of probability models. We need not, however, pursue this internal contradiction in empiricist philosophies of science. It is clear that their aim has been to be as "actualist" as possible.[5]

For the modal realist the analogue of strict causal necessity is a "propensity" existing within the individual system. The exemplar is a single radioactive nucleus. Its probability of one half to decay in a time defined as its "half life" is taken to be a measure of its "causal tendency" to decay in that length of time. Causal tendencies are, for the modal realist, taken to be features of the real world.

Whether macroscopic objects, like human bodies, have propensities, for example, to develop lung cancer, is an open, scientific question. But even if objects like human bodies are strictly deterministic, the number of contributing variables is so large, and the variation among individuals so great, that the similarity between a probability model with probabilities interpreted as propensities and, for example, individual human smokers, is quite strong.

We need not pursue these technicalities any further. My main point is that a modal realist understanding of causality is not tied to nineteenth century conceptions of science.

Laws as Universal Generalizations

The related notions of "law" and "universal generalization" play a major role in logical empiricist and in other accounts of science. In the logical empiricist account of theories, for example, the axioms of a scientific theory were taken to be laws, which were understood as universal generalizations. These notions have a far diminished role in the theory of science unfolding in these pages. In the previous chapter I challenged the view that the laws of motion of classical mechanics function as true, or even well-confirmed, empirical statements. Here I wish to challenge the view that science requires laws that have the form of universal generalizations.

The idea that statements of universal form (All F's are G) play a major role in science goes back to Aristotle. Such statements are well suited to codification in a syllogistic system. Biological classification (species, genus, and so on) still exhibits its debt to the Aristotelian world view.

Although the scientific revolution of the seventeenth century was part of the rejection of an Aristotelian point of view, the goal of discovering universal truths survived. Indeed, that goal was reinforced by the success of Newton's law of universal gravitation. Hume based his analysis of causality on universal laws of association. Kant took universality to be the mark of necessity. And Popper's whole philosophy of science rests on the deductive falsifiability of universal statements by singular statements (This F is not

G). Nevertheless, the importance of universal generalizations is not supported by an examination of contemporary scientific practice.

Let us look at how the venerated law of universal gravitation is treated in standard textbook presentations. Surprisingly, some textbooks of classical mechanics (generally the more advanced) never explicitly state the law of universal gravitation. Nothing of the form "For all bodies. . . ." ever appears. Instead, one finds a treatment of models with two bodies moving in two dimensions subject only to a central force. The inverse square law is introduced as the most important form of a central force. Then one typically finds a derivation of Kepler's laws! These are often mentioned by name. Additional applications include the motion of the planets, the moon, and, recently, artificial satellites. These are cases in which the inverse square force model has been found to fit fairly well. In short, classical mechanics can be presented, and sometimes is presented, without ever invoking gravitation in the form of a universal law.

What, then, is the role of the law in those texts that do explicitly state it? At first sight it seems to be invoked to *justify* the choice of an inverse square force function in constructing theoretical models to solve various problems. A standard problem, for example, is to determine the velocity at which a satellite escapes from its orbit. The solution is derived from a model that incorporates an inverse square force. But one wonders whether the reference to the law of universal gravitation is in fact playing any justificatory role. After the solution to the problem has been derived from the model, it is often pointed out that the value obtained does in fact agree quite well with actual observations. The choice of an inverse square force therefore seems to be justified not by appeal to the law of universal gravitation, but by the fact that the resulting model agrees with observations.

This is not to argue that generalizations play no role in mechanics. I claim only that *universal* generalizations play no role. What one does find is catalogs of cases in which various force functions yield models that fit tolerably well. Putting all those catalogs together yields a truly impressive range of cases. This finite, and not well defined, range of cases constitutes the empirical content of classical mechanics. Nevertheless, the range is surely of sufficient scope to account for the importance of mechanics over the past 300 years. The stature of mechanics is not diminished by eliminating the grand generalizations, which, however important they may have seemed in centuries past, are not essential to the science as it is now conceived.

The idea that the content of a science must include universal generalizations is part of the view that the real content is contained in a set of axioms from which applications are deduced. However much their practice may have deviated from the ideal, Newton and his successors seem to have held this view. So have many philosophers down to the present day. But this ideal does not fit mechanics as it is now taught and understood. The relationship between the basic principles and applications is very different in ways I have tried to describe.

Causal Models and Causal Explanations

Forty years after Hempel's famous paper (Hempel and Oppenheim 1948), explanation continues to be a major topic within the philosophy of science—as evidenced by the recent appearance of two major books (Achinstein 1983; Salmon 1984) and other important works on the topic (van Fraassen 1980, chap. 5; Kitcher 1981). What has changed, however, is the emphasis, particularly in Salmon, on the role of causality in explanation. In this context one might well wonder whether a constructive, causal realism, as part of a cognitive theory of science, has any relevance to recent theories of explanation. The answer is yes, but not in the way one might think.

Although differing with Hempel in major respects, most philosophical writers on explanation agree with his assumption that there exists a distinct category of things properly called "scientific explanations" and that by studying the distinctive features of these explanations one can learn something important about science. As Salmon (1984, ix) put it:

> Our aim is to understand scientific understanding. We secure scientific understanding by providing scientific explanations; thus our main concern will be with the nature of scientific explanation.

Scientific explanation, then, is taken as providing a window on science.

Although everyone from Hempel on has acknowledged the possibility of explaining "laws," by far the major emphasis in philosophical studies has been on explaining *particular events*. This is curious. If there is anywhere one would expect to find scientific explanations, it is in standard textbooks. Yet explanations of particular events are almost nonexistent in textbooks. What one finds, as I have emphasized, is the development of families of models together with exemplary applications to the behavior of particular kinds of systems.

Why, then, do philosophical writers on explanation emphasize the explanation of particular events? Because that is the kind of explanation one finds most often in everyday life. Why did the Chernobyl nuclear reactor break down? What caused the Challenger rocket to explode? And so on. It turns out, therefore, that most philosophical writing on "scientific explanation" is not really about explanations *within* science, but about the use of scientific knowledge in the explanation of events in everyday life. This reflection suggests a very different picture of the relationship between the study of science and the study of explanation than that generally held.

Explaining is a human activity whose practice long antedated the rise of modern science. Indeed, it seems just the kind of activity whose study the cognitive sciences are now poised to undertake. It is already clear that a large part of any cognitive theory of explanation would be an account of how people deploy various sorts of schemata in giving explanations—and in understanding them. And much of this account could be relatively independent of the content of the schemata employed.[6]

From this point of view, all that is distinctive about "scientific" explanations, whether in science or everyday life, is that they deploy models developed in the sciences. Thus, studying scientific explanations is at best a very indirect way of studying science. Little can be learned in this way about science that could not be learned more directly by examining the nature of scientific models and how they are developed.[7]

What science provides for "scientific explanations" is a resource consisting of sets of well-authenticated models. How people deploy those models in the process of constructing or understanding explanations depends on the *extrascientific* context. To take a standard example, it is not part of the science of mechanics to say whether the length of a pendulum explains its period or the period explains its length. What the science of mechanics provides is a model that exhibits the causal structure in which both the length and period are a part. Moreover, the model may even exhibit asymmetries among elements of the causal structure. In the pendulum example length is a free parameter in the original equations of motion—the period is not. The period is derived in the process of obtaining a solution to the equations of motion. But this fact tells us nothing in general about explanatory priority. As van Fraassen eloquently demonstrated with the story of the tower and the shadow (1980, 132–34), one might well explain why a particular pendulum has the length it does by reference to someone's desire to have a pendulum with a predetermined period.

Am I, then, like van Fraassen, advocating a "pragmatic" theory of scientific explanation? Not quite. My position is even more radical. A theory of explanation, for me, is not to be judged by philosophical standards, but by the standards of the cognitive sciences. That is, an empirical theory of explaining would be judged by the sorts of evidence relevant to theories of other higher level cognitive activities such as language comprehension and problem solving. How well van Fraassen's account would fare by these standards is an open question.

Realistic Rejoinders

I will conclude this chapter by replying to four types of arguments against any realistic interpretation of scientific hypotheses. In keeping with the general strategy of this chapter, these arguments are all designed to show that a realistic interpretation is in general incoherent or obviously false. Epistemological arguments against realism will be addressed later.

The Approximation Puzzle

With very little prodding, most scientific realists will admit that, strictly speaking, many, and perhaps even all, realistically interpreted hypotheses are not literally true. But realists will quickly add that most such hypotheses are *approximately* true. At this point the anti-realists, such as Laudan (1981), will reply that what might be meant by "approximately true" it is by no means clear. And they have a point—up to a point.

Approximate truth is not a kind of truth. Indeed, it is a kind of falsehood! Approximately true implies "not exactly true," which means false. Moreover, recent attempts to explicate the notion of approximately truth, such as Popper's notion of verisimilitude (1972, 47–60; Newton-Smith 1981, 52–59), have met with little success. Yet the failure of philosophers to explicate a viable notion of approximate truth must not be taken as grounds for concluding that approximation is not central to the practice of science. Perhaps the source of the difficulty is the philosophers' insistence on understanding approximation in terms of a notion of approximate truth.

Van Fraassen was closer to the mark. He suggested (1980, 9) that for a hypothesis to be approximately correct it must encompass a family of models, one of which is exactly correct, that is, isomorphic with the intended real system. But even this definition is too restrictive. Whatever approximation means in science, it must be true that the dynamical models of classical mechanics are approximately correct for many real systems. Yet if it is agreed that the world is really Einsteinian rather than Newtonian, no dynamical Newtonian model is exactly correct for any real system.

My suggestion, of course, is that the notion of *similarity* between models and real systems provides a much needed resource for understanding approximation in science. For one thing, it eliminates the need for a bastard semantical relationship—approximate truth. For another, it immediately reveals—what talk about approximate truth conceals—that approximation has at least *two* dimensions: approximation in *respects,* and approximation in *degrees.* Armed with just these distinctions, we can begin to attack other recent objections to realism.

THE HISTORICAL ARGUMENT AGAINST REALISM

The historical argument against realism, as popularized, for example, by Laudan (1981a, 1984a), is basically this: The history of science provides many instances of discarded theories whose central terms did not refer to anything—the phlogiston theory and the ether theory being prime examples. Those theories could not, therefore, be regarded as having been even approximately true. Moreover, the evidence on which those theories were originally accepted was no different in kind from that on which our current theories have been accepted. Thus, there is no reason to think that the same fate could not befall our current theories. Indeed, the historical evidence is that many of our current theories will meet similar ends. The historical evidence is strong that realistically interpreted hypotheses often turn out to be false even when claimed to be only "approximately" true.

This argument rests on the unstated intuition that approximation is always a matter of *degrees*. If the ether does not exist, claims involving the ether cannot be just a little bit off. They must be mistaken in some more radical sense. The argument collapses, however, if we abandon talk of approximate truth in favor of similarity between the model and the world, which allows approximation to include respects as well as degrees of similarity.

Whether the ether exists or not, there are many respects in which electromagnetic radiation is like a disturbance in an ether. Ether theories are thus, in this sense, approximations. The fact that there is no ether is one very important respect in which there fails to be a strong similarity between ether models and the world. That failure is a good basis for rejecting ether models, but not for denying *all* realistically understood claims about similarities between ether models and the world.

One way science advances is by discovering new aspects of the world, that is, new respects in which our models might resemble the world. Science also advances by discovering some respects in which similarities between model and world are *not* as commonly thought. Neither sort of advance, however, is inconsistent with constructive realism.

THE SIMILARITY ARGUMENT

At first sight it seems that even a modest constructive realism is incompatible with relativism. Yet both Barnes (1982, chap. 2) and Bloor (1982) have recently presented an argument that, if correct, would place constructive realism snugly in the relativist household. The argument in question is purely theoretical (it makes no appeal to sociological data) and is attributed both to Mary Hesse (1974, chap. 3) and to Kuhn (1961, 1974). Applied to my account of theoretical hypotheses, the argument proceeds as follows.

The relationship between a theoretical model and the world has been presented as one of similarity. Now, as a matter of logic anything is similar to anything else in some respects and to some degrees. To keep theoretical hypotheses from being merely vacuously true, one must specify respects and degrees. But then the truth or falsity of the hypothesis depends entirely upon that specification. And agreement on its truth or falsity depends on a prior agreement regarding specifications of respects and degrees. But these specifications are determined neither by the character of our theoretical models nor by the nature of any real system. If there is agreement, therefore, it can only be because there was a prior agreement on respects and degrees that is socially sanctioned and enforced. Thus, which hypotheses are called true and which false depends entirely on social agreements, which are totally independent of our models and of how the world really is.

I agree with the premises of this argument. A theoretical hypothesis is like a formula with several free parameters. Until the parameters are filled in, the truth or falsity of the hypothesis is indeterminate. Moreover, neither the indeterminate hypothesis itself nor the world determine the respects or the degrees to be included in the claim of similarity. The conclusion of complete social determination, however, does not follow.

The argument assumes that specification of the relevant respects and degrees remains in the background and is not part of the hypothesis itself. The claims whose truth or falsity is at issue therefore merely assert a similarity, with respects and degrees left tacit. In this case the truth or falsity of the explicit assertion of similarity clearly does depend on tacit specifications of respects and degrees. But these specifications need not be tacit, and in science they often are not.

To take a simple sort of case, imagine a model that predicts the values of a specified parameter. A set of measurements are made, all of which are within 10 percent of the predicted values but none of which are within 2 percent. A claim of agreement to within 10 percent would be true, while a claim of agreement to within 2 percent would be false. The relativity to prior agreement has been eliminated by bringing the object of prior agreement within the hypothesis itself. This simple example shows that claims of the kind required by constructive realism can legitimately be made.

The example does not, however, eliminate the problem of deciding whether, in the context, accuracy just within 10 percent is acceptable. That is a separate decision. But the fact that such a decision must be made does not restore the conclusion of complete social determination. Indeed, a judgment on the accuracy of the data must precede the judgment whether that degree of accuracy is acceptable.

We can, however, go beyond this technical refutation of the similarity argument. As is suggested by its more usual formulation in terms of perceptual judgments, the similarity argument is rooted in traditional empiricist epistemology. It assumes a version of the Humean view that there is no natural connection between any two impressions. The only connections are those we impose.

As was pointed out earlier, traditional empiricism is particularly vulnerable to an attack based on post-Darwinian biology. The effect of evolution on our sensory apparatus is known to have been particularly strong. Animals are capable of incredibly fine discriminations among objects in their environment without benefit of social conventions. And so—being fairly intelligent, talking primates—are we. As Bloor (1982) himself acknowledged, no human group among the many that have been studied fails to distinguish red from green, though many do not distinguish pink from red (Berlin and Kay 1969). This fact has a ready explanation in the evolved physiology of our color-sensing mechanisms. Our ability to distinguish colors derives from three pigments sensitive to the whole visible spectrum but which have their peak sensitivities roughly in the blue, green, and yellow regions, respectively. We perceive other colors by adding and subtracting signals from these three types of receptors (P. S. Churchland 1986, 454–55). For at least some perceptual judgments, therefore, the fact of widespread agreement does not require a social explanation. The explanations of evolutionary biology and physiology are sufficient.

Obviously, the acquisition of scientific knowledge, like all human activities, takes place in a social environment. That is not at issue. But humans are also biological creatures with complex, evolved cognitive capacities for interacting with their environment. The task of a theory of science is to discover empirically how all these elements fit together. In this task general sociological arguments that scientific knowledge is completely relative to social agreements are no more helpful than philosophical arguments that scientific knowledge necessarily requires the application of some a priori rules of rationality.

THE PROBLEM OF INDEPENDENT ACCESS

Attempts to describe a realist position within the confines of traditional empiricism have always faced the problem of independent access. Suppose, with Hume or Mill or Russell, that all one "directly" experiences are one's own sensations. How, then, could one compare one's sensations with the

world to determine whether they correspond to it? Since all one could experience is another sensation, one could at best compare two different sensations. In short, there is no "independent access" to the world.

Since radical empiricism is now out of fashion, we no longer find that
argument stated so baldly. Yet it is still influential. A patient reading of
Putnam's recent work (1981, 1983), for example, reveals a more Kantian
version of the argument.

Approaching the problem from the perspective of the cognitive sciences
and evolutionary naturalism allows one to bypass several centuries of fruitless philosophical debate. Rats (Tolman 1948; O'Keefe and Nadel 1978),
and even wasps (Gallistel 1980, 345–49), have the capacity to construct
internal "maps" of their environment. They produce those maps through
causal interaction with the world in a way that yields useful similarities
with that world. Of course, evolution produced the neural capacity for generating such maps—again as a result of long-term causal interactions with
the world. And versions of those same mechanisms exist as well in the human brain (P. S. Churchland 1986).

It would be strange to think that as a result of acquiring greatly enhanced
cognitive capacities, including those for language and self-conscious reflection, humans somehow lost the ability to interact cognitively with the
world in the simpler ways available to lower animals. That is scarcely
possible. The problem lies, rather, with the way humans in various intellectual traditions have represented their own capacities for representing and
interacting with the world. Once the inadequacies of those traditions are
recognized, the way is clear to begin the more fruitful scientific task of
explaining how, beginning with the biological and cultural resources of our
ancestors, we humans have managed to construct modern science.

5

Realism in the Laboratory

The constructive empiricist takes the aim of science to be the creation of models that are empirically adequate, which is to say, models that yield true hypotheses regarding the observable aspects of the world. Theoretical aspects of scientific models may in fact have counterparts in reality, but science cannot justify the belief that they do. And modal aspects of models have no counterparts in reality at all. Yet van Fraassen knows that scientists often speak like realists, even modal realists. They talk about such things as what a proton would do if accelerated in a magnetic field. This just shows, he says, the pervasive character of an empirically adequate model. Our models shape the way we think and talk. Yet this fact concerns only the "pragmatics" of scientific language, not its "logic" or "semantics" (van Fraassen 1980, 199–201).

Though obviously not undertaken for this purpose, recent ethnographic studies of laboratory research provide a microsociological foundation for the constructive empiricist's general account of science. They attempt to show in detail how day-to-day activities in the laboratory generate the apparent reality of theoretical objects. These studies are guided by a "constructivist interpretation of science" that "conceives of the order generated by science as a (material) process of embodiment and incorporation of objects into our language and practices" (Knorr-Cetina 1983, 136).

Advocates of laboratory studies regard their observations in the laboratory as providing empirical support for a constructivist interpretation of science. I propose to counter their claim with a laboratory study of my own. My subjects are nuclear physicists working in a national cyclotron facility. These scientists regard themselves as investigating the structure of the nucleus by bombarding various nuclei with rapidly moving light nuclei, mainly protons, and seeing what comes out. Here I shall not be concerned with their claims about the nucleus, but about protons and neutrons. Any empiricist philosopher or constructivist sociologist would have to count protons and neutrons as "theoretical" or "constructed" entities. My claim will be that the only remotely plausible *scientific* account of what these physicists are doing requires us, as students of the scientific enterprise, to invoke entities with roughly the properties physicists themselves ascribe to protons and neutrons.

If this claim is correct, realism provides the proper account for at least some important areas of modern science. And both empiricism and constructivism will have been shown not to be the generally applicable accounts they have been claimed to be. This finding, of course, will not mean that empiricism or constructivism might not apply in some other areas of science. Which type of account applies must be decided on a case-by-case basis. Only then might one seek some rough generalizations regarding the kinds of circumstances in which realism, rather than empiricism or constructivism, provides the best account.

Contingency and Negotiation

Much of the empirical evidence cited in support of constructivist programs comes from the study of science in laboratory settings. The two best known studies of that kind (Latour and Woolgar 1979; Knorr-Cetina 1981) focused on molecular biologists. What they reported about life in the laboratory can be summed up under the labels "contingency" and "negotiation."

To the outsider a modern scientific laboratory is a chaotic place. People are going every which way doing this and that. At any given time many will be talking to others. Some will be reading, others writing. Some will be fiddling with complex instruments. Others will be at computer terminals. There seems to be nothing remotely resembling the operation of a "scientific method." Indeed, it is difficult to discern any plan or method whatsoever.

Even after identifying a group of people pursuing a single experiment, one finds that it does not proceed even according to the very sketchy plans one can elicit. Scientist X is directing an experiment. Some piece of equipment does not work. X looks around for something else to try in its place. He talks to other scientists. Y suggests he talk to Z who has a likely substitute. Z indeed has the suggested equipment, but she is planning to use it in her own experiment. X negotiates with Z to use the equipment temporarily in exchange for some computer time that Z happens to need right away.

Eventually X's experiment is completed and the results written up to be published. The published paper contains no reference to the complex contingencies and negotiations that actually occurred. This, it is claimed, is one of the ways scientists maintain the appearance that they are "discovering" facts about a preexisting world. In fact, it is concluded, the "world" referred to in the published paper has been "socially constructed" in the laboratory.

Note that it is not merely the *process* of doing an experiment that is claimed to be subject to contingency and negotiation. The claim also applies to the *conclusions* reached. Instances are cited in which scientists, upon gathering some data, spend a great deal of time communicating with other scientists about what it all means. Those interchanges are interpreted as being part of the process of negotiating the conclusion to be published.

Rather than discuss in greater detail cases described by others, I will now describe some of my own observations at the Indiana University Cyclotron Facility. I will confirm with my own examples the findings of contingency and negotiation. What I deny is that these findings support the constructivist program. Contingency and negotiation are compatible with realism.[1]

THE INDIANA UNIVERSITY CYCLOTRON FACILITY

The Indiana University Cyclotron Facility (IUCF) is a national laboratory for research in nuclear physics financed primarily by the National Science Foundation and secondarily by Indiana University. The facility began operation in 1975. It now supports a dozen professors, another dozen Ph.D. associate or staff scientists, several dozen research associates and graduate students, as well as another sixty or so people variously classified as engineers, research technicians, electronics technicians, machinists, cyclotron technicians, operators, and secretaries. The unit's operating budget is on the order of seven million dollars a year. In 1982 the facility acquired its own "Nuclear Theory Group," composed of a half dozen theoretical nuclear physicists.

The majority of the research at IUCF is performed not by the permanent staff but by nuclear physicists from all over the world, though mainly the United States. The IUCF cyclotron occupies a unique niche in nuclear research. It produces a high-resolution proton (or heavier ion) beam in the energy region of 12 to 200 MeV (million electron volts). Scientists wishing to do research with such a source make application to a "Program Advisory Committee" (PAC), which meets roughly every six months. Proposals are judged officially on "scientific merit, suitability to the facility, and prospect for successful completion." In addition to funding, facilities, and support personnel, the main restriction on which and how many experiments can be approved is *time*. What the PAC allots is time measured in eight-hour shifts, twenty-one per week.

CONTINGENCIES IN THE LABORATORY

My observations at IUCF abundantly confirmed the existence of "contingency" and "negotiation" in laboratory research. In describing those observations, I shall employ the language of the physicists themselves. For the most part their language is realistic. In employing this language, I do not intend to be begging any questions against a constructivist interpretation. I shall later offer a separate argument that the use of a realistic idiom is by and large correct. It is the idiom that we, as students of scientific activities, should ourselves employ.

The biggest contingency for researchers is the laboratory itself. Any experiment must be designed around the characterisitics of the beam produced. In addition, experimenters are highly dependent on supporting facilities. For example, one lengthy series of experiments by IUCF scientists required the design and construction of an advanced proton spectrometer. An important factor in the design finally selected was the power of the computers available for data acquisition. Another possible design required much more powerful computers than would be available at IUCF for several years. But since this type of experiment was very new and scientifically exciting, the physicists did not wish to wait for the new computers. It was important for the lab to develop the capability for this type of experiment as soon as possible. And it was important to the careers of the scientists involved to move into this line of research. Besides, since a spectrometer of the alternative design had been used only for higher energy (500 to 800 MeV) protons, the researchers were not certain it would perform better at 200 MeV, despite its more detailed data output (which is why it requires faster computers). It now appears, however, that the next generation of such experiments will be done using the alternate type of spectrometer—which is yet to be built. By then, however, the new computers are expected to be on line, and the IUCF researchers will have had several years of valuable experience.

NEGOTIATION AND EXPERIMENTATION

A case that fits the usual descriptions of "negotiation" occurred one evening halfway through a five-day run. I found the research associate in charge of the experiment at the data acquisition computer in a heated argument with a faculty member. It seemed that the beam had been discovered to be drifting slightly up and down on the target. The argument was over whether this would make any difference to the data being collected. The research associate thought not; the faculty member was not so sure. The consequences of the drift were potentially severe. If the drift did matter, most of the data gathered thus far might be useless. Moreover, since the cause of the drift was not known, finding the source of the problem and correcting it might take several shifts, leaving very little time to collect data before the run had to be completed. The argument, accompanied by several diagrams and equations, continued for an hour. In the end the faculty member gave up and the experiment continued running.

I frankly could not tell whether the faculty member was convinced by the research associate's arguments or simply not sure enough of his own position to recommend shutting down the experiment and trying to stabilize the beam. The main factor may have been that the research associate, not the faculty member, was ultimately responsible for the experiment and would suffer the consequences if the data were later concluded to be useless. The research associate, in turn, may have thought that the chances of correcting the problem in time were so low that he had nothing to lose by continuing and hoping he was right that the drift would not matter. In any case, there was no ready way to check experimentally whether the drift mattered or not. That question could not be answered until more experiments of a similar nature had been performed. That would be months. In the meantime the conclusions one drew from this particular run depended on that prior decision about the possible consequences of a drifting beam.

Without a doubt contingency and negotiation are part and parcel of cyclotron research. But so is realism, as I shall now show.

Producing Protons

Apart from offices, seminar rooms, and workshops, the cyclotron facility has three main areas. One (1 and 2 in figure 5.1) contains the control room and data acquisition areas; the second (3–8 in the figure) is where the beam is produced; the third (11–18 in the figure) is where the experiments are set up. In this section I will concentrate on the area of beam production, which contains the main cyclotron. Here again I shall employ the realistic language of the physicists themselves.

PRODUCING THE BEAM

Proton beams begin as hydrogen gas. Individual molecules of gas are dissociated into two proton-electron pairs, which are subsequently ionized by stripping off the electrons. The protons are then accelerated in an electrostatic generator. The larger of the two generators at IUCF (3 in figure 5.1) produces protons of about 500 KeV (thousand electron volts). The protons

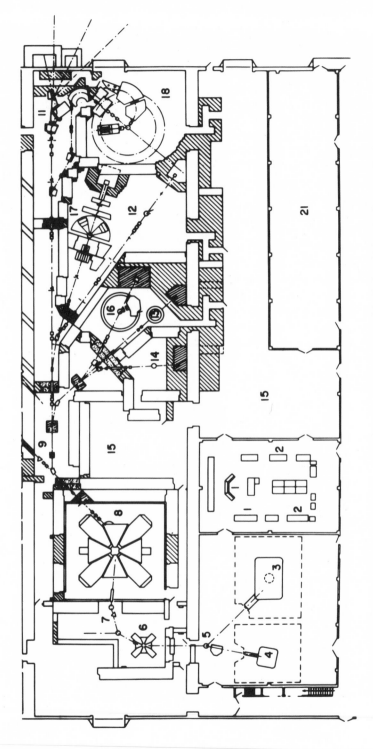

Figure 5.1. The layout of the Indiana University Cyclotron Facility. Reproduced by courtesy of the Indiana University Cyclotron Facility.

emerge from the generator into an evacuated beam pipe that leads across a hallway and through a wall into another room.

Ducking under the beam pipe, moving down the hall a few yards, and turning to the right, brings one to the injector cyclotron (6 in figure 5.1), which is a smaller version of the main cyclotron—in fact, it was originally the prototype for the larger machine. Protons, which enter the injector cyclotron with energies of around 500 KeV, leave with energies of about 15 MeV—a thirtyfold increase. The emerging protons move down the beam pipe, are turned by a powerful magnet, and pass through the wall to the main cyclotron.

Even to an untutored eye the large cyclotron (8 in Figure 5.1) is an impressive piece of machinery. Its four main magnets, each weighing roughly 500 tons, stand 20 feet high. In normal operation it draws 100 kilowatts of electricity and requires 3600 gallons of water per minute to maintain the magnets at a constant temperature of about 85 degrees. This machine typically runs 24 hours a day for several weeks in a row until something breaks or it is shut down for maintenance.

How the cyclotron works can be understood with the help of figure 5.2, which is a schematic cutaway view from above. The operative area, which is evacuated, is only about one inch deep, and so a two-dimensional representation is appropriate. There are only two basic physical principles one must keep in mind. The first is that a charged particle in an *electric* field will experience a force *in the direction of* the field and thus be accelerated in line with the field. The second principle is that a charged particle moving in a *magnetic* field will experience a force *perpendicular to* the field, and thus be deflected in a curved path. Both of these principles had been well established by the nineteenth century.

As one can see in figure 5.2, protons from the injector cyclotron enter at a slight angle from below and are deflected by a combination of magnetic and electric fields into a small orbit near the center of the machine. (Note the steerer magnet, inflection magnet, and electrostatic inflector.) The location and strength of the four large magnets are carefully calculated to turn a moving proton by 90 degrees, so that it moves in a closed orbit that is basically a square with rounded corners. The direction of motion is counterclockwise when viewed from above.

Although not represented in figure 5.2, there are electric fields between magnets A and D and between magnets B and C. These fields are set so as to accelerate protons in the direction of their intital motion. Thus, a proton entering the field of magnet C is moving faster than it was when it left the field of magnet B. Since it is moving faster, it ends up in a slightly larger orbit between magnets C and D than it had when it was coasting between magnets A and B. The same happens between D and A. In this manner, a

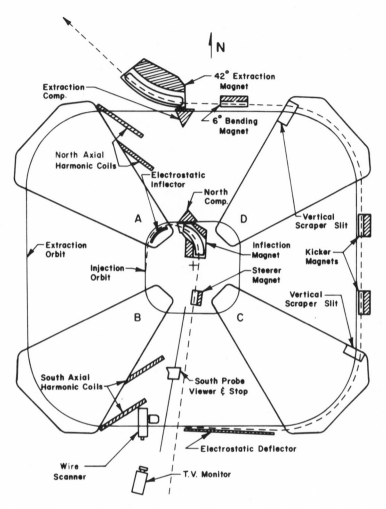

Figure 5.2. A cutaway view of the main stage cyclotron as seen from above. Reproduced by courtesy of the Indiana University Cyclotron Facility.

proton moves in ever larger orbits until it reaches the extraction orbit. This process takes about 600 turns around the machine, at which point the proton has increased the radius of its orbit from roughly 40 to 100 inches. Finally, it is deflected by an electrostatic deflector into a still larger orbit, from which it is extracted by the 42 degree extraction magnet and sent on its way to one of the experimental areas. At this point the proton may be traveling up to one half the speed of light!

CONTROLLING THE BEAM

The amount of control exercised over the operation of the beam is amazing. The most obvious characteristic to be controlled is the energy of the protons produced. By making adjustments all along the three-stage process, operators can maintain the energy to within one tenth of a percent over a range from roughly 20 to 200 MeV.

Even more astounding is the fact that the "polarization" of the beam can be controlled. Protons, like many other particles, exhibit a property that is analogous to the spin of a gyroscope. That property gives these particles a definite orientation in space. It recently became possible to produce protons with a desired spin orientation. But when protons move through a magnetic field, the direction of their spin tends slowly to precess. The protons are therefore gradually reoriented as they are being accelerated in the cyclotron. By the time the protons reach the extraction orbit, they may no longer be pointed in the direction desired for the experiment in progress. One solution to this problem is to extract the protons from a different orbit. Near the extraction orbit, the distance between one orbit and the next is roughly two millimeters. From the control room the exact location of the electrostatic deflector can be set to within a fraction of a millimeter. Thus, if the spin orientation of protons at the 600th turn is not appropriate, the deflector can be moved in to extract the protons at the 599th turn, or the 598th, and so on, until they exhibit a satisfactory spin orientation. And this configuration can usually be maintained for as long as needed.

For a final example of just how much the operators are in control of the beam, look again at figure 5.2. Note in particular the item marked "south probe viewer and stop" between magnets B and C, and note the T.V. monitor just outside the machine at the bottom of the diagram. This monitor is a television camera that is pointed through a glass window directly into the slot in which the protons are moving. What the camera sees appears on a television screen in the control room. The probe, which can be moved along the radius of the orbit by a servomotor controlled from the main control panel, includes a small phosphorescent patch that lights up when struck by a proton beam. The operators can therefore locate the beam within the cyclotron by moving the probe and watching the phosphorescent patch on the television screen. Once the operators locate the beam, they can move the edge of the patch at will in and out of the path of the beam. More importantly, they can redirect the beam ever so slightly by changing the current in various magnets such as the steerer magnet or the inflection magnet pictured in figure 5.2. The operators can watch the spot on the patch move as they make these adjustments. The ability to "see" and manipulate the beam in this fashion is particularly useful when first starting up the machine after maintenance work or a breakdown. The operators can tell at a glance when they have a beam and whether it is in the optimal place. They can even visually distinguish successive orbits near the extraction orbit.

What emerges finally from the main cyclotron is a well-regulated beam of protons. The protons do not, however, form a continuous stream. The electric field that accelerates the protons is not continuous, but alternating at high frequency. This effectively segregates the protons into bunches when they are first injected into the cyclotron. The protons therefore emerge in bursts of roughly 20,000 (20×10^3) protons each. The machine produces about 30 million (30×10^6) bursts per second, which add up to 600 thousand million (6×10^{11}) protons per second. When emerging from the machine, each burst is about six centimeters long, and there are about 600 centimeters between bursts. This works out to roughly 200 watts of power—the power of a large light bulb. That is the basic ingredient for all cyclotron experiments.

Using Protons

Before attempting to draw any general conclusions, I will now describe some aspects of a typical experiment.

The (p,n) Reaction

One type of experiment for which a cyclotron is ideally suited is based on the proton-neutron (p,n) reaction. In this reaction one neutron in a nucleus is replaced by a proton. Figure 5.3 pictures this reaction for the transition from carbon 14 to nitrogen 14. In classical terms the incoming proton undergoes an inelastic collision with a neutron, knocking it out of the nucleus while remaining behind in its place.

The point of most (p,n) experiments is to test various models of nuclear structure. What these models are, and what the experiments reveal about these models, is of no concern here. Our concern at the moment is with what nuclear physicists do with the protons and neutrons. In general terms the answer is easy. They direct protons of known energy and direction at target nuclei. The energy of the expelled neutrons, together with their angle of flight relative to the incoming proton beam, is then determined experimentally.

Figure 5.4 shows the layout of this experiment at IUCF. The target is located in the northwest corner of the building (11 in figure 5.1). The large neutron detectors are located in 16 by 10 by 10 foot huts 50 to 100 yards north of the building.

Now, there are a number of fairly simple questions that anyone survey-ing this experimental setup might ask. These questions do not require any extensive knowledge of nuclear physics. I have discussed the answers with people who actually worked on these experiments, including a person who helped build some of the crucial equipment.

WHY ARE PROTONS AND NEUTRONS TREATED SO DIFFERENTLY?

In these experiments protons are confined to an evacuated beam pipe. In-deed, considerable effort is required to maintain a vacuum in the beam line. The neutrons, on the other hand, fly as much as 100 yards across an open space outside the laboratory. Why this difference?

The answer appeals to the most fundamental properties of protons and neutrons: protons are charged particles, whereas neutrons have no charge. Moving charged particles interact electromagnetically with their surround-ings. Electromagnetic forces are relatively long-distance forces. Thus, a proton shot into the open air, or into a beam pipe that has lost its vacuum, would interact vigorously with air molecules, quickly losing its energy. In-deed, any vacuum leak destroys the beam. Neutrons, on the other hand, because they have no charge, interact with surrounding molecules only by much shorter range, though more powerful, nuclear forces. Roughly speaking, a neutron will interact with air only if it hits the nucleus of an atom making up a molecule of gas in the air. The relative volume of nuclei in any normal volume of air, however, is minuscule. Thus, the average dis-tance a neutron would travel through air before hitting a nucleus (its "mean

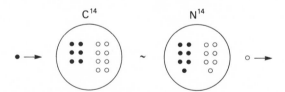

Figure 5.3. A representation of the nuclear reaction p–C^{14}–N^{14}–n.

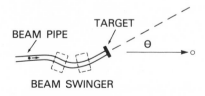

Figure 5.4. An experimental setup for (p,n) experiments at IUCF.

free path" in air) is relatively long. Even traveling 100 yards in air, very few neutrons are deflected from their initial paths.

WHY ARE THE NEUTRON DETECTORS SO FAR AWAY?

Granted, then, that neutrons can travel great distances through air to the remote detectors, what is the point of having the detectors so far away in the first place? In most other experiments performed in this laboratory, the detectors are placed very near the targets.

As noted above, one of the quantities measured in (p,n) experiments is the energy of the ejected neutron. Because neutrons have no charge and thus interact little with other particles, it is difficult to measure the energy of moving neutrons. One can measure their energy indirectly by having them interact with other particles, such as protons, and then measuring the energy of the secondary particles. But such indirect methods introduce considerable uncertainty into the measurement.

This experiment measures the energy of the neutron more directly by determining its *time of flight* from the target to the detector. From the neutron's time of flight researchers can easily calculate its average velocity and thus its average energy. The detector for such measurements need signal only the presence of a neutron; it does not have to measure its energy. All that is needed in addition is some way of knowing when the neutron left the target. This is accomplished by timing when the bursts of *protons* hit the target. By electronically screening out three out of every four bursts of protons, one can be quite sure which burst of protons produced a given neutron that has been detected and thus determine the time between when the protons hit the target and when the ejected neutron reached the detector.

Now, for any calculation of average velocity based on measurements of time and distance traveled, the larger the distance the greater the accuracy. The reason the detectors are so far away, then, is to achieve sufficient accuracy in the measurements of the neutron energy.

It is interesting to note that the time-of-flight measurements reveal just what anyone would expect. For fixed proton energy and a fixed direction of the ejected neutron, the farther away the neutron detector, the longer the time of flight. Similarly, everything else being equal, the greater the velocity of the incoming proton, the shorter the time of flight of the neutron. Again, everything else being equal, the time of flight increases for greater angles between the incoming proton and the ejected neutron. In these respects protons and neutrons are similar to billiard balls.

Why Are the Neutron Detectors So Large?

In most cyclotron experiments the detectors are kept as small as possible to make accurate determinations of the angles of ejected particles. In this experiment the neutron detectors are relatively large, with a surface measuring roughly one by three feet. Why are these detectors so large?

Consider a hemisphere behind the target whose radius is equal to the distance to the detector. The further away the detector, the smaller its area relative to the area of the whole hemisphere and thus the smaller the chance that it will be hit by any particular ejected neutron. But one must detect a large number of neutrons so that the variance in the measured time of flight will be low. Given a relatively low rate of (p,n) reactions, even at maximum beam intensity, one must do something to increase the count rate, because beam time costs roughly $1,000 an hour. The solution was to make the distant counters larger, thus sacrificing some accuracy in the measurement of the angle, to achieve satisfactory accuracy and efficiency in the determination of neutron energies.

What Is the Role of the Beam Swinger?

Finally, what is a beam swinger? (See figure 5.4.) And what role does it play in these experiments?

Recall that besides the energy of the expelled neutron, the other quantity to be measured is its angle of flight relative to the direction of the incoming proton beam. In most cyclotron experiments the corresponding angles are measured simply by moving the detectors around the target. In these (p,n) experiments, however, the detectors, together with their supporting electronics, are so large that they cannot easily be swung around from one angle to another in the course of a series of measurements. What the researchers do, therefore, is leave the detectors fixed and change the angle of incidence of the incoming proton beam. That is the purpose of the beam swinger. It consists of two large magnets that change the angle at which the beam strikes the target. It literally swings the beam out of its initial path and then back to the target at a predetermined angle. The detector that originally measured the presence of neutrons moving directly in line with the proton beam now measures those moving off at the angle determined by the setting of the beam swinger.

To the best of my knowledge IUCF is the only place in the world that does time-of-flight (p,n) experiments using a beam swinger. At least one other laboratory does do this *type* of (p,n) experiment, but it uses an entirely different setup to vary the angle between the incident protons and the ejected neutrons.

Experimentation and Realism

There can be no doubt that the nuclear physicists I have observed are realists in the sense that they believe something is going round and round in cyclotrons, down the beam pipes, and striking the targets. Moreover, they believe this something has roughly the properties ascribed to protons— mass, charge, momentum, and so forth. This belief is obvious not only from the scientists' words, but also from their actions. They have routinely told me, for example, that protons make only about 100 orbits in the small injector cyclotron as opposed to roughly 600 in the large machine. And they have built in elaborate safety mechanisms, such as alarms and interlocking doors, to make sure no one gets in the way of a proton beam or the radiation it produces. These are not the words or actions of people who harbor any serious doubts about the reality of protons.

But then few students of the scientific life would deny that most scientists are realists regarding many subjects of scientific inquiry. The question is whether scientists are *correct* in so believing, and whether we, as students of the scientific enterprise, should adopt their accounts as even roughly correct descriptions of what they are doing.

Empiricist philosophers, such as van Fraassen, would argue that these scientists are not really justified in believing that there are such things as protons. Constructivist sociologists would claim that these scientists, through their social practices, have deceived themselves into thinking that their own social constructs have an independent existence. My view is that these scientists are more or less corrrect in their beliefs about protons and neutrons and that we, as interpreters of science, must invoke the reality of things like protons and neutrons if we are to provide an adequate scientific account of their activities.

Let us be clear about the state of the dialectic. Protons and neutrons are paradigm cases of "theoretical" or "constructed" entities. For empiricists or constructivists to admit even a minimal (constructive) realist interpretation in this case would be to give up the applicability of their account to a central area of modern science. On the other hand, constructive realists need not insist that everything claimed by some scientific group to exist does in fact exist. Nor need they deny that there are some cases in which an empiricist or constructivist account does apply. It is just that those accounts do not apply to this central case.

Producing and Using Protons

My argument is simple. The only remotely plausible, generally scientific account of what is going on at the cyclotron facility is the one I have already given, or something very much like it. These nuclear physicists are *producing* protons with desired characteristics, such as energy, and then *using* them, together with other particles, to investigate the properties of various nuclei. To say that they are "producing" and "using" protons implies that protons exist.[2]

The immediate response of both empiricists and constructivists will be that my description of the situation begs the question. If these physicists are producing and using proton beams, then of course protons exist. But are they really producing and using proton beams?

Here it is important to distinguish the correctness of the judgment from an analysis of the grounds for that judgment. I think it is simply undeniable that these nuclear physicists are producing and using particles with roughly the properties ascribed to protons. It is not, of course, undeniable in the Cartesian sense that doubt would be self-contradictory. But it is undeniable in the more ordinary sense that one could not doubt that there are people, computers, or large magnets in that laboratory. The judgment about protons seems to me one whose correctness we should take as a basis for further explanations of what is going on in the laboratory. The judgment is not itself problematic.[3]

My basis for this claim is my experience in this laboratory, together, no doubt, with my earlier training in physics. I have tried to describe some of that experience. My description, of course, is fragmentary, and it will mean more to someone with a similar background and similar experiences than to others. But that is unavoidable.

None of this means that we cannot say more about the basis for the judgment that these physicists are producing and using protons. But our *analysis* of this basis cannot be taken as undermining the judgment itself.

Manipulation and Control

Empiricists and constructivists focus almost exclusively on what scientists say or what anyone can observe with unaided senses. My focus is on scientists' physical interactions with the world. What we see in a cyclotron facility is a great deal of physical manipulation and control of particles like protons. I have no doubt that nuclear physicists' ability to manipulate protons in so many detailed ways has much to do with their unquestioned realistic attitude toward protons. And it is largely by observing the exercise of this ability that we, as outsiders, can tell that their realistic attitude is correct.

One may wish to view my position as an application of some more general principle such as: "Whatever can be physically manipulated and controlled is real." Such a principle would serve to describe my position, but it could not legitimately be taken as providing useful support for the position. It is far too vague, and it probably could not be made much more specific without inviting obvious counterexamples. For example, the nature and extent of the manipulation and control required to conclude that something is real is obviously important. It would be impossible, however, to specify in general terms what type and how much manipulation is required. This must be decided on a case-by-case basis. Here I would claim that the types of manipulations I have described as occurring in the cyclotron facility are obviously sufficient. Nuclear physicists can literally make protons do loops on command.

Nor can one infer great detail from a general ability to manipulate and control. What we can be sure of is physicists are producing and using things with properties *like* those ascribed to protons. There can be no guarantee that things that can be manipulated in many ways are *exactly* as they are currently conceived to be. They may always turn out to be somewhat different. Protons may be composed of three quarks. If the quark theory is correct, protons are not "elementary particles," as they were earlier conceived. Nevertheless, it will have to be true that three quarks bound together to make up a proton behave much like protons have been thought to behave in medium-energy nuclear research. They may do these things somewhat differently, and they may do many other subtle things besides. But neither possibility negates the original, realist claim.

From Theoretical Entity to Research Tool

The manipulation of protons in cyclotron research is obviously not done for its own sake. For the most part physicists produce protons to investigate the structure of the nucleus. It is the details of nuclear structure that count as "theoretical" in the context of nuclear research. Nuclear physicists do often question whether particular aspects of proposed models of nuclear structure have any counterparts in reality. But they never ask such questions about protons. In this context, then, protons function as research tools, not theoretical entities.

It was not always this way, of course. Between 1907 and 1919, when Ernest Rutherford and his colleagues were conducting their famous experiments with alpha particles at Manchester, protons were indeed among the most theoretical of entities. The existence and the properties of protons were then the object of investigation in the way that the structure of the nucleus is now the object of investigation. Since then, however, physicists have succeeded in learning a great deal more about protons. Of equal importance, they have developed an elaborate technology that makes it possible now to employ protons as tools in contemporary research. The development of the cyclotron, beginning in the 1930s, is an obvious example of just such a technological innovation.

Empiricists and constructivists are forced to downplay the physicists' distinction between research tools and the objects of current study. For them protons and nuclei are equally theoretical or constructed. For them the fact that some theoretical or constructed entities play a role in the investigation of other theoretical or constructed entities is just that—a fact. Focusing on contemporary research tools that were once theoretical entities exposes this empirical failing of both empiricism and constructivism.

Why Not Be Content with Scientists' Beliefs?

One might grant that nuclear physicists are realists about protons, and that they are so because of their great ability to manipulate and control proton beams in their everyday research. Still, one might ask, why should we, as outsiders seeking to understand the activity of modern science, invoke the reality of protons in our own explanations of what scientists do? Why not be content to claim that scientists have these beliefs and act on them in their day-to-day research?

The answer is that such a restriction severely impoverishes the explanations of scientific activity one can give. It limits the resources used in giving explanations to those of everyday life. For example, I explained the fact that neutron detectors are placed at a great distance from the target in terms of the need for accurate time-of-flight measurements of the ejected neutrons. An explanation couched solely in terms of the beliefs of physicists about the behavior of neutrons would not only be cumbersome, it would be substantially weaker. Why restrict oneself to giving weaker explanations when one could easily give stronger ones?

Constructivists have no qualms about assuming the reality of *other people*. They are perfectly willing to explain Jones's actions by reference to a conversation Jones had with Smith. Is that not assuming too much? Should we not rather say that Jones believed he had a conversation with Smith? Surely that would be silly. Restricting explanations of physicists' activities to invoking only their beliefs about protons, rather than protons themselves, is just as silly.

The Limitations of Empiricism

I will confine my remarks here to van Fraassen's liberal form of empiricism since any criticisms of this form apply with greater force to more extreme forms. My general conclusion is that, regarded as an empirical theory of science, van Fraassen's account is itself not empirically adequate. It fails as an empirical theory of science even when judged by its own standards.

THE OBSERVABLE AND THE THEORETICAL

All forms of empiricism, including van Fraassen's, conflate the distinction between the observable and nonobservable with the distinction between the nontheoretical and theoretical. The result is a single distinction between observable and theoretical. Probably the most telling, and most often told, criticism of empiricism is that this conflated distinction cannot be made in a nonarbitrary way (Suppe 1974). In particular, and more important, it cannot be made in a way that provides any understanding of the actual practice of science.

My observations in the laboratory provide a vivid example of this standard criticism. Physicists do make a distinction between "observation" and "theory" or, as I would prefer to say, between "data" and "model". But that is not the distinction empiricism requires. In (p,n) experiments, for example, the *data* are the energy of the incoming protons together with the energy and angle of the outgoing neutrons. The *models* involve details of nuclear structure. The data are used as the basis for decisions about the models. This is roughly how physicists describe the situation, both informally and in their publications.

Empiricists must reject this description as a correct account of the situation. For them the data must be something observable like counter readings. The energy of the ejected neutrons is for them just as theoretical as the details of nuclear structure. No matter that the actual counter readings never appear as "descriptions of the data" in any scientific accounts.

Indeed, in current research what empiricists would call data turn out to be numbers or graphs produced on computer terminals or printouts. There are no longer any counters to read. The signals from the detectors go directly into a computer. If one were to take seriously the empiricists' view that the sole empirical content of science, "what science is all about," is what humans can observe, one should have to conclude that much of modern science is about what is written on computer printouts. That should give any empiricist cause for reflection.

MODELS AND RESEARCH TOOLS

Van Fraassen, of course, claims that the actual existence of theoretical entities makes no difference to the practice of science and, in any case, there is no way to justify beliefs about such entities. He thus makes no fundamental distinction between the status of physicists' models of protons and the models of the nucleus currently under investigation. The accumulated observational evidence for the empirical adequacy of proton models is simply much greater now than it once was.

Consider the operator at the control panel of the cyclotron. The experimenters come over and request that the energy of the beam be increased to 180 MeV for the next series of measurements. The operator gives the appropriate signals to the control computer and shortly receives back the expected information that the energy of the beam is now at 180 MeV.

Van Fraassen would view this exercise as itself an experiment. Its success is only additional evidence for the empirical adequacy of our models of the proton and the cyclotron together with all its related instrumentation. All anyone can claim to know is that if one gives these instructions to the control computer, one gets back the expected response. No one has legitimate grounds for believing that there really are such things as protons. By implication, no one has grounds for thinking that he or anyone else is producing, manipulating, or otherwise causally interacting with protons.

Whatever plausibility this account has derives from focusing on a single isolated action, like a particular adjustment of the beam energy. The plausibility vanishes when we consider the hundreds of such adjustments made every day, the process of designing and performing hundreds of experiments, the process of designing, building, and maintaining a cyclotron facility over a period of years, and so on and on. As a result of this continuing and varied experience, contemporary nuclear physicists never even think about questioning the existence of protons or wonder whether they have adequate evidence for their beliefs about protons. It is only the difference in scale relative to our natural perceptual apparatus, and the consequent need for elaborate technology to mediate the interaction, that tempts contemporary philosophers to question whether physicists' beliefs about protons are justified.

I must reemphasize that my claims here apply only to things like protons that have achieved the role of research tools. I make no such claims, for example, about the detailed structure of the nucleus, which is an object of current research. That we might at least sometimes properly adopt a realist account in these sorts of cases requires additional argument.

THE PHENOMENOLOGY OF SCIENTIFIC ACTIVITY

There is one final rejoinder that an empiricist like van Fraassen would make. He would grant that my descriptions of activities in the laboratory are the "appropriate" descriptions. That is, what I have presented may indeed be a correct description of the "phenomenology of scientific activity" (van Fraassen 1980, 80–83). The theoretical, and even the modal, aspects of the models scientists use become part of their "world picture." Perhaps they cannot help treating their models this way.

Nevertheless, we should understand scientists as operating under a "supposition" that this is what the world is like. As examples of suppositions van Fraassen would cite the familiar practice of reductio proofs in logic and mathematics, where one begins with the supposition that a proposition is true and proceeds to show that it leads to a contradiction. Another example is a play. Within the context of a play we can distinguish truths from falsehoods. It is true, *in the play,* that the butler did it and false that the wife did it. But this is all only within the supposition of the play.[4]

This reply suffers from the philosophers' preoccupation with language to the exclusion of causal interaction with the world. But let us suppose it can be extended to activities as well. It then appears as a clever attempt to preserve the empirical adequacy of an empiricist account of science. It provides an empiricist description of what we see and hear in the laboratory.

The trouble is that this response is totally ad hoc, and in the end vacuous. Except for the desire to save an empiricist account of science, there is no reason in the world to think that activities in the laboratory are part of a grand supposition. If that were true, then, indeed, all the world is a stage. All our lives are nothing but complex sets of suppositions. That is a strange conclusion from someone who would "deliver us from metaphysics" (van Fraassen 1980, 69).

The Limitations of Constructivism

Constructivists examine the microsociological processes that result in the publication of research findings. Along the way they find much uncertainty and controversy that is eventually resolved through a complex, and highly contingent, process of social interaction. The result, it is claimed, is more a social construction than a description of nature. But the influence of the process of social interaction is not recorded in the final presentation of new findings. And so it appears that the scientist is describing nature pure and simple when in fact the voice of nature is barely audible, if it can be heard at all.

By focusing on the process of ongoing research, constructivist investigations have made valuable contributions to our understanding of science as it is currently practiced. But an excessive concern with current scientific conclusions, especially those in highly controversial areas, makes the constructivist account seem better than it is. It misses what scientists take for granted, namely, the role of previous findings in current research, particularly findings that have made possible the instrumentation for current research. There was a time, before 1920, when one could have argued that protons are a social construct. But that time is long past. Today protons must be regarded as being no less real than protozoa. Quarks, perhaps, are still up for grabs (Pickering 1984). To maintain a constructivist stance regarding protons requires extreme maneuvers indeed.

THE AZANDE GAMBIT

In direct response to comments by me on the importance of instrumentation in scientific research, Karen Knorr-Cetina replied:

> If I were Feyerabend I would now draw a comparison to the magic of the Azande and say that all belief systems have an elaborate machinery to substantiate and sustain their beliefs. . . . All belief systems in a way try to externalize . . . their beliefs to attribute them, to have nature speak to them, or to have their gods speak to them, for example, through some chicken oracle, rather than to have their own opinions or their neighbors' opinions registered and taken seriously. And some of these oracles, for example, are highly sophisticated and articulated. One could ask the question whether science is not just our magic.

Although she immediately went on to say, "I am not Feyerabend. . . . I won't go quite as far," the point as stated is sufficiently common in the literature to deserve comment.[5]

If the reference to witchcraft were merely part of an attempt to illuminate the nature of science, that could be helpful. But more often the reference to Azande practices seems a disguised way of insisting that to be able to distinguish science from witchcraft, one must produce some noncircular, a priori, criterion of demarcation. But that is no more necessary for a theory of science than is an a priori justification of induction.

More positively, the reference to Azande practices is intended to illustrate the general thesis that a society's image of the *natural* world is completely on a par with its image of the *social* world. Both are culturally relative, and in neither case is it possible to prove one image superior to the other. But this is just a posture. If our goal is to understand the natural world in a way that makes modern technology possible, we simply must admit that contemporary scientific practice is superior to Azande witchcraft. The task for the cultural study of science is to explain why this is so, not to deny the obvious. On the other hand, no part of any such explanation should claim that our science is superior because *we* are more rational. Nor does it follow that any "primitive" peoples would be irrational not to adopt the goals of science and employ scientific means to those goals. On these points the cultural relativists are certainly correct.[6]

CONSTRUCTIVISM AND EMPIRICISM

Probably the most appealing constructivist reply to the argument of this chapter is to claim that my descriptions of nuclear physicists as producing and using protons are simply question-begging. All that I have described is simply people doing things like turning knobs and watching bright spots on TV screens. The rest is interpretation. This response, however, leads straight to empiricism.

Consider the hypothetical case of a laboratory very much like a cyclotron facility except that the particles being accelerated are not protons but BBs, shot pellets. Unlike protons BBs can be seen with the naked eye, held in the hand, and rolled between the fingers. Now suppose that my descriptions of scientists producing and using high-energy BBs are analogous to my descriptions of activities in the cyclotron facility. Suppose further, however, that for some reason I never actually see a BB myself. Nevertheless, I conclude that these scientists are producing, using, and otherwise causally interacting with BBs.

Would my conclusion be question-begging? Would I myself actually have to see or feel a BB to prevent my conclusion from being question-begging? To insist that I must would be to embrace a fairly strong form of empiricism. The only basis for allowing my conclusion for BBs but not for protons would be the empiricist objection that no one can handle protons the way one can handle a BB. The constructivist, therefore, is presented with the dilemma of either giving up the claim that constructivism provides the correct interpretation of this case or explicitly allying constructivism with empiricism.

Now, there is a radical way out of this dilemma. That is to insist that BBs are every bit as much socially constructed entities as protons—despite the fact that one *can* roll them between one's fingers. The trouble with this way out is that it would make everything a socially constructed entity. As a result, the constructivist account of science would cease to be a scientific claim based on the empirical study of science. It would become, rather, a simple consequence of the general, metaphysical thesis that everything is socially constructed. The excursions by constructivists into the laboratory would serve not to test a constructivist model against the actual practice of science, but merely to attempt to interpret the practice of science in terms of the constructivist model.

My view, of course, is that the constructivist model cannot be made to fit the practice of experimental nuclear physicists. I infer that it cannot be made to fit many other scientific fields either, including some, such as molecular biology, studied by prominent constructivists.[7]

Geometrical Cognition in Nuclear Research

Many philosophers and sociologists of science have been suspicious of claims about theoretical aspects of reality. From a cognitive viewpoint the discovery of aspects of reality beyond ordinary sense experience is a remarkable achievement of human ingenuity. The major problem is to *explain* how such an achievement has come about. In concluding this chapter I will examine two directions one might explore in search of such an explanation. One turns inward toward the neuroscientific bases of specialized cognitive abilities. A second turns outward toward the technology that is so prominent in contemporary scientific research.

GEOMETRY IN THE LABORATORY

One of the many striking features of a cyclotron facility is the geometrical layout both of the laboratory itself and of individual experiments. These geometrical features of the facility are to some extent evident in figures 5.1 and 5.4. I was particularly surprised to see that part of the geometry of the (p,n) experiments, the paths leading to the large neutron detectors, is clearly visible in aerial photographs taken while these experiments were in progress! Various geometrical aspects of cyclotron research appear also as diagrams in published papers, in formal presentations, and in informal conversation. Just as many people cannot talk without gesturing, experimental nuclear physicists cannot discuss their work without drawing pictures. Is the prominence of geometrical relationships just an accidental feature of cyclotron research, or is it somehow important to understanding why the research is so successful? Recent work in animal behavior and the neurosciences suggests it may be very important.

Cognitive Mapping

On general evolutionary grounds one would expect that mammals, being self-mobile, would have evolved very good mechanisms for representing motion in space and time. Most everything mammals do—fighting, fleeing, foraging, reproducing, and so on—requires coordinated movement. This obvious evolutionary point is reflected in some recent behavioral and neurobiological research.

Figure 5.5 shows the layout of a now classic form of animal experiment (Morris 1983). The large circle represents a tank of water roughly four feet in diameter and just deep enough so that a laboratory rat placed in the tank cannot touch the bottom. It is forced to swim. The water is colored with milk, both to make it opaque and to mask odors. It is also cool, roughly 25 degrees centigrade, to provide the rat with ample motivation for seeking an escape. The dotted rectangle represents a clear plastic platform resting on the bottom with its top just below the surface. The landmarks placed around the tank are ones known to be clearly distinguishable by rats.

One standard form of the experiment proceeds as follows. A naive rat is placed in the water facing the edge of the tank at a randomly selected position around the circumference. It typically swims in roughly decreasing circles until it encounters the platform, whereupon it climbs onto the platform to escape the cool water. It then sits there surveying its surroundings. Shortly thereafter, say 30 seconds, the rat is picked up and reinserted in the tank at a different place on the circumference. This time it does not swim in circles but, after briefly surveying its surroundings, heads off in the general direction of the platform. After four or five trials the rat will swim almost directly to the platform from any location on the circumference. That the

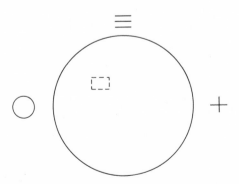

Figure 5.5. An experimental setup for rat navigation experiments.

rats are not detecting the platform by smell or other direct cues is shown by moving the platform on a subsequent trial. The rat goes to where the platform was, is very confused when it does not find it, and only reluctantly goes into something like its original circular search pattern.

To perform this task, the rat's cognitive system must somehow be able to appreciate spatial relationships. That is, on the first trial it learns the location of the platform relative to the surrounding landmarks. Then, when placed back into the tank, it must somehow determine the location of the platform relative to its *new* position to be able to head in the right direction. How does it do this?

One current theory is that when it first locates the platform the rat constructs a cognitive map, that is, an internal representation of the spatial features of its surroundings. When placed back into the tank at a different position, it locates itself in its cognitive map and then heads in the direction of the platform as indicated by the map. The rat's brain thus is not literally doing geometrical calculations, any more than does a person reading a map. The representational system, the map, already contains all the necessary information. It is an additional virtue of this theory that it postulates a direct connection between the representational system and the sensorimotor system, that is, between thought and action.

Recent experiments have even been interpreted as showing that at least part of the mapmaking mechanism is located in the hippocampus (O'Keefe and Nadel 1978; O'Keefe 1983). Recordings of the firing rate of single cells in the hippocampus of freely moving rats show that specialized cells (now called place cells) fire differentially as a function of location in the environment. In effect these cells are like lights on a map that indicate "you are here." In different environments the same cell may be assigned different locations with no apparent relationships among maps for different environments. Nevertheless, rats are capable of retaining at least a half dozen different maps simultaneously.

Moreover, rats who have mastered the tank revert to their original circular search pattern after an operation that destroys the hippocampus. Control rats given sham operations retain their ability to find the platform. Apparently, destroying the hippocampus makes it impossible for the rat to use its previously constructed map (Morris 1983).

Finally, some recent theories of sensorimotor control suggest that the brain works more on geometrical than on algebraic principles (P. S. Churchland 1986, chap. 10). Studies of the cerebellum (Pellionisz and Llinas 1982) indicate that one of its functions is to transform sensory input vectors into motor-controlling output vectors. Information from the inner ear, for example, regulates muscle control to keep one upright while walking. One recent theory (P. M. Churchland 1986) suggests that sensory modalities, such as vision and audition, may be represented in a neuronal array that is sandwiched together with other arrays representing real space (a "map") and muscle control. The resulting neuronal "sandwich" provides a direct way of connecting locations in visual and auditory space, for example, with positions in action space, as when one reaches for a ringing telephone.

From Rats to Researchers

What has this line of research to do with nuclear physics? Just this. It shows that humans, by virtue of their biological evolution, have a highly developed capacity to represent spatial relationships. This capacity is located in the preverbal parts of the brain, closely connected with the motor control system. The overwhelming tendency among experimental nuclear physicists to think and communicate in terms of diagrams suggests that they are tapping these sorts of preverbal cognitive and sensorimotor capacities. The action involved in drawing a diagram, for example, is just the sort of thing that would engage such capacities. Thus, there is good evidence that humans are naturally endowed with the kinds of capacities necessary to conceptualize and carry out cyclotron experiments. That would provide at least part of an explanation of why they are so good at it.

Once one begins to think in these terms, applications to other sciences immediately leap to mind. Perhaps part of the reason why molecular biology advanced so rapidly after the structure of DNA was discovered was simply because humans are very well adapted to thinking in terms of geometrical structures. By this hypothesis it is not just that genetics was reduced to chemistry, but to *structural* chemistry, that accounts for some of the success of molecular biology.[8]

Visualizability and Quantum Theory

When quantum theory was in its infancy (1900–1925), there was much talk among both physicists and philosophers about "visualizability." Quantum mechanics, they said, did not admit of the sort of easily visualizable models familiar from classical mechanics. The success of quantum theory showed that visualizability was an inessential feature of physics. The real content of the theory, its commentators concluded, is in the equations, not the visual images that one might have. In this, as in many related matters, Einstein dissented. He always claimed to think in terms of nonverbal imagery. Yet long since those debates have been forgotten, we find nuclear physicists using visualizable models, indeed, mostly ones that can readily be drawn on paper. What does this fact imply for our understanding of contemporary nuclear research?

One implication is that the study of nuclear research reinforces the view that scientists operate with clusters of models—and not with implicit formal deductive systems. In this case it is obvious that the various models are not all compatible. Some nuclear models, of course, are defined by the equations of quantum theory. Others are straightforwardly classical, as when we picture the (p,n) reaction as an inelastic collision between billiard balls. Still others are semiclassical, as when we picture protons as spinning solid spheres with quantized angular momentum. Moreover, physicists well recognize the virtues and limitations of the various models and thus seldom make mistakes because they are using an inappropriate model.

Now, it is fairly clear from my observations in the cyclotron laboratory that during their day-to-day work, *experimental* nuclear scientists prefer classical or semiclassical models to the more complex quantum mechanical ones. Moreover, they almost universally describe the more classical models as being more "intuitive" than the quantum ones. My explanation for these facts would be that working in the laboratory context requires constant use of one's sensorimotor and preverbal representational systems. Classical and semiclassical models can presumably be processed at the preverbal level of cognitive mapping. The full-fledged quantum theoretical models, by contrast, require more involvement with verbal or symbolic representational systems. Operating these systems requires, I presume, more concentration as well as more training.

Suppose we grant, then, that the success of cyclotron research is partly explained by the fact that experimental nuclear physicists are using their hippocampuses (and other neural structures) to good advantage. It is still a long way from using these cognitive capacities to being able to use protons as projectiles. Our cognitive mapping mechanism, after all, evolved in the context of the macroscopic world of small mammals, not in the microscopic world of contemporary nuclear science. What fills the gap? Theory? Only partly. The main connector between our evolved cognitive capacities and the micro world of nuclear physics is *technology*.

The Role of Technology in Scientific Research

My first visit to the Indiana University Cyclotron Facility evoked many memories. Among the strongest were ones having little to do with nuclear physics. I thought of the automobile assembly plant I had visited as a schoolboy and of a more recent visit to the generating plant beneath the Hoover Dam. What triggered those memories were sights, sounds, and even odors. The sight of large machines beside which their creators and operators seem insignificant. Banks of amplifiers. The continuous hum of large transformers. The phut, phut, phut of vacuum pumps. And the smell of oil and ozone.

The overwhelming presence of machines and instrumentation must be one of the most salient features of the modern scientific laboratory. Yet contemporary historians, philosophers, or sociologists of science rarely mention these things. I will forgo the temptation to speculate on the reasons for their neglect. The fact is that it utterly distorts the picture of science that emerges. The development of science depends at least as much on new machines as it does on new ideas.[9]

Neglect of the role technology plays within science is fairly recent. Among the most enduring images of science is that of Galileo with his telescope. A commonly held, but now often overlooked, view has been that instruments are man-made extensions of the senses. What humans could not see with the unaided eye, Galileo could see with the help of his telescope. My view is that this tradition captured an important insight. It is technology that provides the connection between our evolved sensory capacities and the world of science.

A TECHNOLOGICAL FIX FOR THE DUHEM-QUINE PROBLEM

Pierre Duhem (1914) was the first to emphasize that we cannot test a model directly against our observations. Instead, the model must be embedded in an experimental context that is defined, in part, by other models. It is only in the total experimental context that the data have any significance. The problem raised by this view, Duhem argued, is that the importance of negative evidence becomes ambiguous. Negative evidence indicates that something is wrong, but it does not show unambiguously that the model under test is mistaken. The failure of the observations to come out as predicted may be because one or more of the *other* models defining the experimental context do not fit the situation at hand. The conclusion often drawn from this observation is that the model under examination may always be retained in the face of apparently negative evidence simply by suitable modifications in the other models employed.

Among more recent philosophers Quine (1953) has done the most to popularize Duhem's view. Duhem's problem, in Quine's hands, became one of the foundations for Quine's "holistic" picture of scientific knowledge. For him and most other modern philosophers of science, Duhem's problem is a consequence of a very simple logical schema. Let H represent the hypothesis being tested, A the required auxiliary hypotheses, and O

the predicted observations. Elementary logic then provides the following schema:

$(H \,\&\, A)$ imply O. Not-O. Thus, Not-H or Not-A.

In short, the import of the negative observation, Not-O, is that either H or A is false; but there is no telling which is false—or indeed whether both are false. This formulation of the problem obviously puts H and A on the same logical footing. Both are "hypotheses" that equally might be false. Though a staple of philosophers, the Duhem-Quine problem has also recently been invoked by constructivist sociologists (Knorr-Cetina and Mulkay 1983) in support of their own position.

Part of my earlier criticism of both constructivism and empiricism was that they assign a similar status to questions about the reality of protons and questions about the reality of nuclear structure. But protons, I argued, are not nearly so problematic as the detailed structure of the nucleus. Protons are research tools, which is to say, they are part of the technology employed in investigating nuclear structure. My technological solution to the Duhem-Quine problem follows similar lines. Scientists' knowledge of the technology used in experimentation is far more reliable than their knowledge of the subject matter of their experiments.

Much of the technology in the cyclotron facility, for example, was literally built by people in the lab. This is true of most frontier research. The scientist is doing things that have never been done before, and that requires much equipment that did not previously exist. That equipment cannot be bought; it has to be custom-made. This is why machinists and electronics experts are part of the permanent staff at the cyclotron facility. Obviously, not everything is built on-site. Some things can be purchased from scientific supply houses. Some of those things, like amplifiers, are manufactured. Others are built more or less by hand, one at a time. In either case there is little question that the operation of those instruments is exceedingly well known. We are therefore not dealing here with "hypotheses" on anything like the scale of the models being tested.

Of course, it does not follow that failure to get an expected experimental result is never due to malfunctioning equipment. Indeed, such failures are all too common. But they can be discovered by fairly well-established means, though even that process is not always easy. The folklore at the cyclotron facility includes a story about the time an entire research team spent a full shift (eight hours) looking for the possible source of aberrant data only to discover that they were the result of a burnt out pilot light on an amplifier! Someone had foolishly wired the pilot light in series with a functioning component.

TECHNOLOGY AS EMBODIED KNOWLEDGE

In spite of the Duhem-Quine problem several philosophers of science have recently argued for the importance of "background knowledge" in science (Boyd 1981; Shapere 1984). This argument is well placed, indeed, long overdue. But here too the tendency is to think of background knowledge exclusively in theoretical terms as hypotheses previously confirmed—that is, as *propositional knowledge*. This assumption ignores the importance of technology within scientific research. At least some background knowledge is better thought of as *embodied knowledge*. It is embodied in the technology used in performing experiments.

The cyclotron, for example, was designed with the intention of using it to accelerate protons (among other things). The design may therefore be thought of as embodying some of our knowledge about protons, such as their charge and mass. When the first cyclotrons were built in the 1930s, no one gave any thought to the possibility of producing beams of spin-polarized protons. Now our knowledge of how to produce those beams has been built into the design of modern cyclotrons and their accompanying instrumentation.

TECHNOLOGY AND SCIENTIFIC PROGRESS

Kuhn's writings focused attention on the question whether there is such a thing as progress in science, and, if so, how to characterize it. In the ensuing debate commentators, for example, Laudan (1977), have generally assumed that the question should be addressed at the level of theory or research programs.

Now it seems to me fairly easy, in fact, to characterize several dimensions of theoretical progress. For example, our models become more detailed over time as we discover new aspects of the world. And the range of real systems covered by our models increases. But thinking of the technology used the laboratory as embodied knowledge allows us to characterize an altogether different dimension of scientific progress.

The proton was once among the most theoretical of particles. Scientists had real questions about the reality of any such thing. Now the proton has been tamed and harnessed to the equipment used to investigate other particles and structures: quarks, gluons, and the shell model of the nucleus. Thus, some of what we learn today becomes embodied in the research tools of tomorrow. That is undeniable progress of a very different, and very important kind.

6

Scientific Judgment

Earlier I described a restricted form of realism, constructive realism, which, I claimed, is conceptually coherent and neither vacuous nor obviously false. I then attempted to show that this realist model fits a central scientific case, the use of protons and neutrons as probes in experimental nuclear physics. This demonstration bypassed all the standard philosophical arguments that attempt to prove that realistically interpreted hypotheses, even if they make sense, could not possibly be justified. I shall examine several of these arguments later in this chapter.

The main problem to be addressed here is different. Not all the theoretical models currently accepted as roughly correct by the members of a scientific field will be embodied in existing experimental technology. This fact raises the question how these *other* theoretical models come to be generally accepted. What is desired is a naturalistic account consonant with current thinking in the cognitive sciences.

Within the cognitive sciences our problem would come under the general heading of human judgment. And, indeed, some human judgment research will be relevant to our concerns. But right from the start I should like to narrow the focus to a special case of human judgment, namely, *decision making*. Our topic is thus not scientific judgment in general, but scientific decision making.[1]

Scientists as Decision Makers

My main reason for narrowing the focus from judgment to decision making is that there exists a highly developed set of concepts and principles for thinking about decision making. No similarly well-developed concepts exist for dealing with judgment in general. Yet, although most students of the scientific enterprise, including historians and sociologists, speak naturally of scientists as "choosing" a theory or "deciding" that some model is correct, philosophers are just about the only group ever to have taken seriously the idea of applying decision theoretic concepts to the phenomenon of theory choice in science. Why is that so?

The vast majority of the literature on decision theory focuses on what is explicitly called "normative" decision theory. The literature is therefore part of the general program of characterizing and justifying a model of *rational choice*. The reason for this focus is to be found in the historical development of decision theory out of the theory of games. That theory was originally part of mathematical economics, and economists, like philosophers, have traditionally been concerned primarily with rational agents.[2]

On the other hand, everyone acknowledges at least the possibility of using decision theoretic models to *describe* actual decision making by real people. Although relatively neglected by economists and philosophers, this descriptive approach is just the kind of thing that could be of use in the empirical study of science, whether by historians, philosophers, sociologists, or cognitive scientists.

Once one begins to think of scientists as decision makers, one realizes that the typical scientist makes dozens of decisions of all sorts every day. From this perspective a decision to *accept* a model as providing a roughly correct picture of some part of the world has to be a rare event. Nevertheless, because this type of decision is so important in science, and because it has received so much attention by all students of scientific life, I shall concentrate on it here.

One other type of decision, however, requires at least some attention. This is the decision to *pursue* a type of model. Pursuit is obviously one thing for theoreticians and a different thing for experimentalists. Theoreticians pursue a family of models by elaborating their formal structure. Experimentalists pursue models by designing and executing experiments to test their applicability to phenomena in the world. Though perhaps more common than the decision to accept or reject a model, the decision to pursue a model is still a relatively rare event. But it is also a very important event in the life of any scientist.

For those readers unfamiliar with decision theory, I shall begin with an elementary exposition of the basic structure of simple decision models. The desired *interpretation* of these models will be introduced as required. Only later will I attempt to *identify* particular instances of decision making that might fit these models.

Basic Decision Models

Most accounts of decision making, whether descriptive or normative, share a common core: a model of the *structure* of decisions. Although exceedingly simple, this model is extraordinarily useful.

The basic model includes an *agent* and a *world*. The agent is assumed capable of various *actions,* including the action of choosing among alternative courses of action, called *options.* For there to be a nontrivial decision the agent must have available at least two different options. Without options there are no decisions to make.

The world, in this model of decision making, is as it is. And what happens to the agent depends on how the world is. For the decision to be of any importance, it is assumed that what happens to the agent depends also on which option the agent chooses. If the result of the choice is the same no matter what the agent does, the decision problem is again trivialized.

Next the model assumes the agent is not omniscient. In particular, the agent does not know the actual state of the world. The model does, however, assume that the agent can characterize at least two different *possible* states of the world. Thus, although the agent does not know in what state the world actually *is,* it does know of several states in which the world might be.

This simple model of decision making imposes important constraints on both the set of options available to the agent and the assumed possible states of the world. Both sets must be formulated to be simultaneously *exclusive* and *exhaustive.* That is, the model assumes the agent is committed to choosing one and only one of the options. Moreover, whatever the option chosen, the agent ends up in one and only one of the designated states of the world.

The structure of the simplest sort of nontrivial decision allowed by this model is exhibited by the matrix in figure 6.1. This structure makes precise the notion of the outcome of choosing a particular option. An outcome is an option-state pair. That is, it is having chosen a specific option in a world characterized by one of the possible states. Thus, the decision characterized in a two-by-two matrix has four possible outcomes. In general, a decision with m options and n possible states will have $m \times n$ possible outcomes.

THE ROLE OF VALUES IN DECISION MAKING

The skeleton of a decision model depicted in figure 6.1 is of very limited applicability without adding an explicit component for *values.* The theory

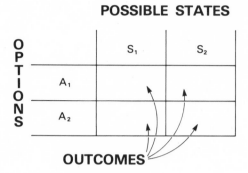

Figure 6.1. The basic structure of a decision.

assumes, therefore, that in making decisions agents seek to further their own interests, goals, or values. In this model the value component is expressed as a relative evaluation of the *outcomes*. That is, the agent's values do not attach directly either to options or to states of the world, but to action-state pairs. What is valued is having chosen a particular option in a definite state of the world. As far as the agent is concerned, the only result that matters is obtaining the value associated with a particular outcome.

The evaluation of outcomes may take several different forms. One simple form is a linear rank ordering. That is, in more technical terms, the ranking is closed and transitive. *Closure* means that every outcome is ranked. *Transitivity* means that if outcome *A* is ranked greater than or equal to outcome *B,* and *B* greater than or equal to *C,* then *A* must be ranked greater than or equal to *C.* A slightly more restrictive form of valuation would eliminate the possibility of ties or equal rankings; that is, it would require that every outcome have a different rank.

It is interesting to note that not even *normative* decision theory dictates how the agent should evaluate specific outcomes. Rather, it imposes only general constraints, like linear ordering, on the agent's relative evaluation of outcomes. Thus, even normative decision theory basically takes the agent's own values as given.

Extensions of the Basic Model

The model of decision making presented thus far is relatively uncontroversial, but also seriously incomplete. It does not, for example, provide any structure for incorporating new information of the kind that might result from experiments. To complete the model, however, takes us into disputed territory.

I shall now present an extension of the basic model that is currently popular among philosophers and decision theorists—the Bayesian model of inference and decision making (Jeffrey 1965, 1983; Rosenkrantz 1977; Winkler 1972). I shall then consider whether a Bayesian account might adequately explain how scientists choose models. My conclusion will be that it does not, for the solidly empirical reason that experiments indicate otherwise.

Bayesian Decision Models

Bayesian models can be developed as a natural extension of the basic decision model represented in figure 6.1. These models postulate a considerably enriched structure for the agent's preferences for the various outcomes. In particular, a Bayesian agent's preference structure is said to be *coherent*. That is, a Bayesian agent is *defined* as an agent with a coherent preference structure. The nature of such coherent preference structures is usually characterized axiomatically. Here I will proceed informally.[3]

Among the seemingly innocuous requirements is that the Bayesian agent's ordering of preferences for the possible outcomes be *transitive*. Yet even this reasonable-sounding requirement is often violated in everyday life, as in the familiar children's game "Scissors, Paper, Rock"—scissors cut paper, paper covers rock, but rock smashes scissors.

The requirements that end up doing most of the work depend on the notion of a *probabilistic mixture* of outcomes. This is a kind of lottery in which the agent chooses between one outcome with probability p and another outcome with probability $1 - p$. Thus, instead of being presented with two definite outcomes, the agent has, so to speak, a ticket in a lottery whose prizes, with probabilities p and $1 - p$, are the designated outcomes.

Imagine an agent faced with three possible outcomes labeled, in descending order of value, 1, 2, and 3. One of the crucial axioms of the Bayesian model requires that for some probability strictly between zero and one the agent would prefer the probabilistic mixture of outcomes 1 and 3 to the certainty of outcome 2 by itself. This is not a trivial requirement. Indeed, versions of this requirement have been debated since the time of Pascal (Hacking 1975).

To bring out the import of this axiom, suppose the agent is an ordinary person like you and the three outcomes are (1) a long, healthy life; (2) a long, healthy life marred only by a mild chronic condition such as a slight tendency to ulcers; and (3) a long life of unrelenting, miserable pain. It is a direct consequence of the axiom in question that there is some probability strictly less than one for which you would prefer the probabilistic mixture of 1 and 3 to the certainty of 2 by itself. Can this be you? Would you not prefer the certainty of the mild condition to any finite chance, no matter how close to zero, of a long, miserable life? Why risk suffering something terrible just for a chance, no matter how near to one, to achieve a slight improvement in one's state?

Why, then, do Bayesian theorists postulate such a complex, and in some ways counterintuitive, structure to an agent's preferences? Because the complexity is necessary to achieve the desired solution to what is generally regarded as the fundamental problem of decision theory. The problem is to characterize a mathematical function that picks out the rationally preferred option solely as a function of the agent's preference structure. That the axioms in question do characterize the desired function is a matter of strict, mathematical proof.

It is an immediate theorem that the agent's preferences for the outcomes can be characterized by a utility function that has the following two properties. First, the agent's utility for outcome 1 is greater than that for outcome 2 if and only if the agent prefers 1 to 2. Second, if the agent is indifferent to the choice between outcome 2 and a given probabilistic mixture of 1 and 3, then the utility assigned to 2 equals the weighted sum of the utilities of 1 and 3, that is, p times the utility of 1 plus $1 - p$ times the utility of 2.

A second theorem proves that a Bayesian agent's attitudes toward the possible states of the world can be characterized by a probability function. This function is then interpreted as representing the agent's *degrees of belief* in the actuality of the various possible states of the world.

Given a utility function for outcomes and a probability function for possible states of the world, we can define the agent's expected utility for any option. This is the weighted sum of the utilities of the outcomes associated with that option, the weights being the probabilities associated with the corresponding states.

Finally, it follows mathematically that the Bayesian agent prefers that option with the greatest expected utility. In other words, the rational choice for a Bayesian agent turns out to be that option which maximizes its expected utility. Operationally, we might imagine that the agent fixes its preferences in a coherent fashion, calculates the expected utility of each option using its derived utility and probability functions, and then chooses the option with the greatest expected utility.

That one can mathematically solve the fundamental decision problem in such an interesting and powerful manner explains much of the fascination with coherent preference structures on the part of mathematicians, economists, and analytic philosophers. Still, we must eventually come to terms with the question whether such models apply to real people, including real scientists.

THE ROLE OF NEW INFORMATION

Up to this point there is nothing in the basic model of decision making, or in its Bayesian extension, that requires the agent to be a scientist, or even an individual person. It might equally well be a group, perhaps a whole scientific field. But let us now be somewhat more specific and interpret the agent as being an individual scientist, say an experimental nuclear physicist.

Let us assume, furthermore, that our scientist is considering two different, well-defined models of some nuclear process. One is a "standard" model, S, and the other a "new" model, D. The options facing the scientist may thus be interpreted as being: (A) accept the standard model as being the best representation of the actual process, or (B) accept the new model as the better representation of the actual process. The two possible states of the world, then, are that the standard model better represents the actual process or that the new model does. The resulting decision problem is represented in figure 6.2.

It is easy to see that the simple two-by-two matrix of figure 6.2 might well fail to capture the actual decision situation. The resulting Bayesian decision model implies that our scientist will be predisposed to choose one or the other of the two models even in the absence of any evidence. A Bayesian scientist must always have coherent preferences regarding the various possible outcomes. Thus, unless the expected utilities of the two choices just happened to be exactly the same, a Bayesian scientist could simply choose the option with the greatest expected utility and forget about gathering any evidence.

But a real scientist might well claim not to have made any decision as to which model best captures the nuclear process in question. Indeed, it might be claimed that the purpose of current experiments is to get information that will help decide the issue.

We can easily resolve this situation simply by introducing a *third* option,

	S FITS BEST	D FITS BEST
CHOOSE S		
CHOOSE D		

Figure 6.2. A representation of a scientists's decision problem in choosing between two models.

which is to suspend judgment on the choice between the first two options. In fact, we could easily imagine that this third option has the greatest expected utility for a Bayesian scientist. But how then does new evidence enter the decision problem so that one of the two substantive options might later have the greatest expected value? It does so by modifying the Bayesian scientist's probability function.

BAYES'S THEOREM

To see how new evidence may alter the decision problem, we need only extend the agent's probability function slightly to include various possible states of evidence in addition to the possible states of the world. We can then define the *conditional probability*, $P(S/E)$, of the state in which S is the correct model, relative to evidence E. This is interpreted as the scientist's degree of belief that S represents the actual state of the world, given evidence E. The Bayesian scientist's probability function necessarily includes the following version of the Reverend Thomas Bayes's famous theorem:

$$P(S/E) = P(S) \times P(E/S)/P(E),$$

where $P(S)$ represents the agent's initial degree of belief, or *prior* probability that S is the true state of the world. The posterior probability, $P(S/E)$, becomes the prior probability for later processing of additional data. The accumulation of scientific evidence is thus reflected in the Bayesian scientist's evolving probabilities for the possible states of the world. Eventually, the expected utilities of the options might change sufficiently that the Bayesian scientist would be led to give up suspension of judgment and choose one of the proposed models as being correct.

BAYESIAN INFORMATION MODELS

Many philosophical advocates of Bayesian methods, such as Jeffrey (1956, 1965, 1985), prefer *not* to think of scientists as choosing models in the sense of deciding that one particular model fits the world better than any rival. The reason is that once a person has decided that a particular model is correct, it seems that person is committed to treating the model as correct in any situation whatsoever. Yet the person knows full well that the chosen model might not be correct, and indeed, earlier did assign a positive probability to its not being correct. Thus, it would be possible for an agent to find itself in a new situation for which maximizing its expected utility using the original probabilities would lead to one choice, while simply accepting the hypothesis as true (in effect assigning it a probability of one) would lead to another.

In general, then, it is claimed that scientists, acting in their role as scientists, ought not to choose models, but simply assign probabilities to the corresponding states of the world. These probabilities can then be used, either by the scientists themselves or by others in making practical decisions. I will refer to such agents as *Bayesian information processors.*

Note, however, that the Bayesian information-processing model does not escape assuming that its agents have the highly structured preferences of Bayesian decision makers. Even in the information-processing model scientists do make decisions, and when doing so they maximize their expected utility. What Bayesian information processors do not do is make decisions about the truth of hypotheses themselves.

One might object that the original argument against making decisions about the models themselves presumes much too simple a view of the relationship between pure and applied science. It could well be that scientists in the context of "pure" science do provisionally "decide" which model is correct, but when moving to an "applied" context resurrect the relevant probabilities and proceed to maximize their expected utility just as the Bayesian information processor does.

Whether scientists, as scientists, *should* be Bayesian information processors has been much debated by philosophers of science (Levi 1967, 1980; Kyburg 1974; Giere 1976b, 1977). We need not pursue this debate any further, for there is now overwhelming empirical evidence that no Bayesian model fits the thoughts or actions of real scientists. For too long philosophers have debated how scientists ought to judge hypotheses in glaring ignorance of how scientists in fact judge hypotheses. Before presuming to give advice on how something ought to be done, one should first find out how it is done. Maybe it is now being done better than one thinks. Indeed, attempting to follow the proffered advice might be detrimental to scientific progress.

Are Scientists Bayesian Agents?

Widespread contemporary interest in Bayesian models began with the publication of L. J. Savage's *Foundations of Statistics* (1954). By the early 1960s psychologists had begun investigating the application of Bayesian models to real people. That research has continued, with increasing intensity, up to the present. The overwhelming conclusion is that humans are not Bayesian agents.

THE BOOKBAG AND POKER CHIP PARADIGM

The earliest experimental paradigm was that of Ward Edwards and his associates at the University of Michigan (Phillips and Edwards 1966). The subjects were undergraduate male students, and the test involved bookbags containing a thousand red and blue poker chips. In a typical experiment there were two bags, one with a 70–30 ratio in favor of red and one with the same ratio favoring blue. One bag was selected by flipping a fair coin. The subjects were briefed to make sure they understood that this meant that the odds were 50–50 that the bag they had selected was the one with red chips predominating. Then began a series of "samplings" from the chosen bag, one chip at a time with replacement. To control the data available to the subjects, the researchers did not allow them to sample the bags themselves, but instead gave them signals in the form of flashing red or blue lights representing a sampled chip. After receiving a bit of "data," the subjects updated their assessment of the probability they had the bag with mainly red chips by distributing 100 washers between two posts, one for each of the two possible types of book bag. They of course began with 50 washers on each post.

A typical result is as follows. After obtaining an apparently random sequence of 12 chips containing 8 red and 4 blue, subjects' estimates of the probability they were sampling the predominantely red bag rose to between .7 and .8. The probability calculated using Bayes's theorem is .97. From published reports it seems that none of several hundred subjects ever gave an estimate that high.

Figure 6.3 shows results for a number of trials with a typical subject. The horizontal axis records the difference between red and blue in the sample. The vertical axis shows the subject's estimate of the ratio of the probability of having the red-predominant bag to having the blue-predominant bag. The solid line represents this ratio calculated for a Bayesian information processor. It is interesting to note that the subject's estimates of the probability ratio, like those of the ideal Bayesian agent, are roughly linear with the difference between red and blue. Only the slope is smaller. This indicates that human subjects are in fact fairly systematic in their non-Bayesian behavior.

This sort of experiment has been tried in many variations, with similar results. The conclusion originally drawn from these experiments was not that humans are not Bayesian agents, but that they are systematically *conservative* in their use of data to revise their initial probability judgments. That is, relative to an idealized Bayesian agent, human subjects do not revise their judgments as much as the data warrant. But more recent investigators have drawn a more radical conclusion: "In his evaluation of evidence, man is apparently not a conservative Bayesian: he is not Bayesian at all" (Kahneman, Slovic, and Tversky 1982, 46).

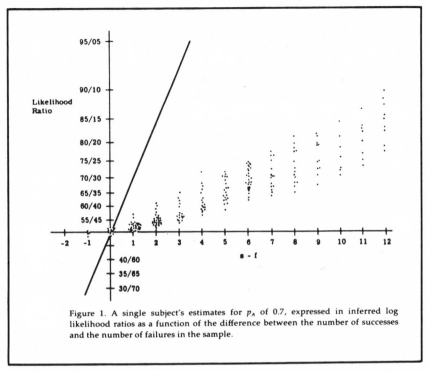

Figure 1. A single subject's estimates for p_A of 0.7, expressed in inferred log likelihood ratios as a function of the difference between the number of successes and the number of failures in the sample.

Figure 6.3. The responses of a single subject to random samples from a binomial population with a 70–30 success-to-failure ratio. Reproduced from Edwards (1968).

VIOLATION OF THE CONJUNCTION RULE

Edwards (1968) recounted that he and his colleagues had originally thought that failures to fit the Bayesian model would be fairly difficult to demonstrate. They were surprised to find so clear and systematic a divergence from the Bayesian model in such a relatively simple context as sampling poker chips from a bookbag. Yet substantial deviations from the Bayesian model have now been demonstrated even in the simplest possible judgments of probabilities for compound events.

Kahneman and Tversky presented the following problem to a variety of subjects including (1) statistically naive undergraduates, (2) statistically aware graduate students in psychology and education, and (3) statistically very sophisticated graduate students in the decision science program at Stanford Business School (Kahneman, Slovic, and Tversky 1982, 90–96):

Linda is 31 years old, single, outspoken, and very bright. She majored in philosophy. As a student, she was deeply concerned with issues of discrimination and social justice, and also participated in anti-nuclear demonstrations.

Please rank the following statements by their probability, using 1 for the most probable and 8 for the least probable.

a. Linda is a teacher in elementary school. (5.2)
b. Linda works in a bookstore and takes Yoga classes. (3.3)
c. Linda is active in the feminist movement. (2.1)
d. Linda is a psychiatric social worker. (3.1)
e. Linda is a member of the League of Women Voters. (5.4)
f. Linda is a bank teller. (6.2)
g. Linda is an insurance salesperson. (6.4)
h. Linda is a bank teller and is active in the feminist movement. (4.1)

The numbers in parentheses are the mean rank assigned to the answer by all subjects.

One would think that everyone would be aware that the probability of two events' happening together cannot be greater than that of either one by itself. And it is, of course, an immediate consequence of the axioms of probability that for any two events, A and B,

$$P(A \ \& \ B) = P(A) \times P(B/A) = P(B) \times P(A/B).$$

Thus, the probability of the conjunction mathematically cannot be greater than the probability of either conjunct by itself.

Yet the average ranking of the conjunctive event "Linda is a bank teller and is active in the feminist movement" by the subjects in this study is obviously higher than that of one of its conjuncts, "Linda is a bank teller." Moreover, at least 85 percent of all subjects judged the conjunctive event more probable than its conjunct. And perhaps most surprising of all, there were no statistically significant differences in responses among the three groups of subjects. Even having taken advanced courses in probability and statistics seemed not to matter in this experiment.

Kahneman and Tversky have an explanation for these results. It is that in making this sort of judgment, people rely on a "representativeness heuristic." That is, they judge the likelihood of the possible situations presented by how representative they think the person in the original description is of the class of people depicted. And judgments of representativeness have a structure different from that of probability judgments. In particular, adding more detail to the description of a possible situation can make it *more representative* of a class of situations, even though the added detail necessarily makes the given situation *less probable*.

This explanation was tested by asking a group of naive subjects to rank the eight situations "by the degree to which Linda resembles the typical member of that class." The average rankings were almost identical to the probability rankings.

This is just one of many experiments of this type. Violation of the conjunction rule, whatever its explanation, is well established. Human beings are not naturally Bayesian information processors. And even considerable familiarity with probabilistic models seems not generally sufficient to overcome the natural judgment mechanisms, whatever they might be.

MEDICAL DIAGNOSIS

One might expect that experienced scientists operating in their own field of expertise are better Bayesians than naive subjects or even sophisticated subjects presented with everyday examples. But this assumption apparently is not so. In an experiment at Harvard Medical School, 20 fourth-year medical students, 20 residents, and 20 attending physicians were asked the following question in hallway interviews (Casscells et al. 1978, 999):

> If a test to detect a disease whose prevalence is 1/1,000 has a false positive rate of 5 percent, what is the chance that a person found to have a positive result actually has the disease, assuming that you know nothing about the person's symptoms or signs.

Before going any further, readers are invited to give their own response simply in the form: high (over 75 percent); medium (around 50 percent); low (below 25 percent).

Answers ranged from 0.095 percent to 99 percent. Almost half (27/60) of the respondents gave the answer as 95 percent. The average response was 56 percent. Only 11 of the 60 participants—four students, three residents, and four attending physicians—gave the correct answer, which is 2 percent! The answer is based on the assumption, which the experimenters seem to have taken for granted, that the test has a false negative rate of zero, that is, it never fails to detect a person who in fact has the disease. With this assumption, the correct answer is provided by a simple application of Bayes's theorem.

Figure 6.4 diagrams an account of the situation that is more graphic than an algebraic computation. The diagnostic test in fact sorts the population

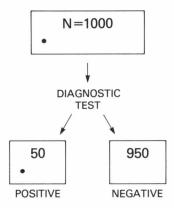

Figure 6.4. A representation of a diagnostic test with false positives of 5 percent.

into two groups, those with a positive response and those with a negative response. Because of the false positive rate of 5 percent, the expectation is that 50 out of every 1,000 disease-free members of the population will end up in the positive group. The other 950 will end up in the negative group. If it is assumed that there are no false negatives, the one person with the disease will end up in the positive group. The test therefore does perform an important diagnostic function. Before the test one's chance of picking out a person with the disease by random selection was only 1/1000. After screening, one need only sample the positive group. Here one's chance of finding a diseased person by random sampling is 1/50. The screening has increased one's chances of picking out a diseased person by a factor of 20. That is a big increase. But not as big as most people think.

This experiment is standardly cited as an example of a natural human tendency to ignore the *base rate* when judging probabilities. In this case, when considering whether a person who tested positive has the disease, one tends to ignore the information that only 1/1000 have it to start with. The fact that this base rate is relatively small compared to the false positive rate of 1/20 leads to a *grossly* mistaken estimate.

THE BASE-RATE FALLACY

In the best known experiment demonstrating the neglect of base rates, first performed by Kahneman and Tversky in 1972 and since repeated many times by themselves and others, subjects were presented with the following problem (Kahneman, Slovic, and Tversky 1982, 156–57):

A cab was involved in a hit and run accident at night. Two cab companies, the Green and the Blue, operate in the city. You are given the following information:

(a) 85% of the cabs in the city are Green and 15% are Blue.

(b) A witness identified the cab as Blue. The court tested the reliability of the witness under the same circumstances that existed on the night of the accident and concluded that the witness correctly identified each one of the two colors 80% of the time and failed 20% of the time.

What is the probability that the cab involved in the accident was Blue rather than Green?

In the many experiments employing variations on this question, the answer given most often is 80 percent, which corresponds to the reliability of the witness. The average answer is roughly the same. The "correct" answer, calculated using Bayes's theorem with equal initial probabilities for an accident by any particular cab, is .41. The credibility of the witness is offset by the low base rate for Blue cabs. But typical subjects ignore the base rate and make their judgment solely on the information about the witness.

A very interesting variation on the above experiment substitutes the following information for (a):

(a') Although the two companies are roughly equal in size, 85 percent of cab accidents involve Green cabs and 15 percent involve Blue cabs.

This formulation carries exactly the same base rate information as the earlier version. Yet individual answers to this problem vary greatly, and the base rate is generally no longer ignored. The average answer is around 60 percent, which is a compromise between the correct answer and the reliability of the witness.

The standard interpretation of this latter result is that people are more apt to make some use of a base rate if it can be fit into a *causal schema*. Here the obvious schema is just the vague idea that there is something about the Green cabs, or their drivers, that is causally responsible for their having more accidents than the Blue cabs.

CAUSAL SCHEMATA

The importance of causal schemata in everyday reasoning has been verified in a wide range of experiments. Many of these inquiries focus not on statistical reasoning, but on the more general phenomenon of causal attribution, particularly to human agents (Nisbett and Ross 1980). Here we will focus on a scientific example.

The following problem (among others) was given to 165 undergraduate students at the University of Oregon:

> Which of the following events is more probable?
> (a) That a girl has blue eyes if her mother has blue eyes.
> (b) That the mother has blue eyes if her daughter has blue eyes.
> (c) Neither. The two are equally probable.

Somewhat fewer than half of the subjects gave the "correct" answer, which is that the two are equally probable. Among those giving another answer, three times as many chose the answer corresponding to the normal causal direction—from mother to daughter (Kahneman, Slovic, and Tversky 1982, 118–20).

The standard interpretation of this type of experiment distinguishes *causal* reasoning (cause to effect) from *diagnostic* reasoning (effect to cause). People seem to be naturally more comfortable with causal reasoning, and to place more confidence in it. Even when the information conveyed is the same in both directions, subjects favor causal over diagnostic reasoning.

Bayesian information models, of course, consider only the information conveyed and so ignore the difference between causal and diagnostic reasoning. We therefore have here a very important respect in which humans fail to fit a Bayesian information model.

SCIENTISTS ARE NOT BAYESIAN AGENTS

The typical modern scientist is not overtly a Bayesian agent. Nuclear physicists, for example, never publish estimates of the odds of one hypothesis being true relative to others. Nor is this just a result of the enforced style of science journals. Even in their informal moments, nuclear scientists do not talk like Bayesians. The claim of those who would use Bayesian models descriptively must be that scientists are in fact "intuitive Bayesians," whatever they may say, and whatever methodology they might espouse. It is the claim that typical scientists are intuitive Bayesians that is refuted by the empirical studies described above.

One way to deny this refutation is to insist that the subjects studied in these experiments are not representative of scientists as a whole. But that is a very difficult tack to maintain. Many of these experiments have involved subjects with advanced scientific training—even many with considerable experience in using probability and statistical methods. Why should physicists, geologists, and biologists be any different? Indeed, education in the physical sciences generally involves *no* courses in statistics, statistical inference, or sampling. Thus, even though none of these experiments seems to have been performed using practicing physical scientists as subjects, there is no reason to suspect that they would perform much differently.

Note that this is not to deny that physical scientists understand probability theory and use it effectively. Nuclear physicists, for example, are quite accomplished at performing probabilistic calculations in the context of statistical mechanics or quantum theory. At issue is only whether they use a probabilistic framework in organizing their own judgments concerning the various theoretical hypotheses involved in their research. Here the evidence says that they do not.

A number of recent writers have attempted to reconstruct *historical* cases along Bayesian lines (Dorling 1972; Rosenkrantz 1977, 1980; Howson and Franklin 1985; Franklin 1986). And those efforts were apparently meant to provide an explanation of why scientists made the choices they did make. In the light of the above research, those claims must be either mistaken or beside the point. Scientists, as a matter of empirical fact, are not Bayesian agents. Reconstructions of actual scientific episodes along Bayesian lines can at most show that a Bayesian agent would have reached similar conclusions to those in fact reached by actual scientists. Any such reconstruction provides no explanation of what actually happened. For that we need another account.

Satisficing Models

Rejecting the Bayesian account of theory choice does not mean abandoning all decision theoretic accounts of how scientists choose one model over another. It simply means that we shall have to employ some descriptively more adequate way of filling out the basic decision model developed earlier. A good candidate for a better model was developed 40 years ago by Herbert A. Simon in the context of business administration (Simon 1945, 1957, 1979, 1983; March and Simon 1958).

Bounded Rationality

Simon began by considering a model of an ideally rational agent like that used in classical economics. Such an agent is ideally rational in two respects. First, it knows all the options physically (or even logically) open to it, and it knows all physically (or even logically) possible states of the world that might be relevant to its decision. Second, like a Bayesian agent, it has a coherent preference structure. It is therefore able to calculate its expected utility for each option and thereby determine the option with the maximum expected utility. The rational agent of classical economics is a *maximizer.*

Simon's strategy for understanding human organizations was to look at how they structure the decisions of individuals within the organization. But he found the requirements placed on "economic man" too severe. "Administrative man" cannot be fully rational as required by the standard economic models. To understand administrative man, Simon developed a model of an administrative agent that operates under conditions of what he came to call *bounded rationality.* Its rationality is bounded in just those respects that characterize rational economic agents.

First, administrative agents operate with a very restricted set of options and possible states of the world. By and large, those are the options and states presented by the agents' immediate situation. Second, administrative agents are seldom able to construct a coherent preference structure for the possible outcomes of their decision problems. Nor are they generally capable of calculating an expected utility for each outcome. They can, however, distinguish those outcomes that are "satisfactory" from those which are not. In some cases they may be able to rank order all outcomes. Rarely can they define a cardinal utility function over the outcomes.

In short, unlike classical economic agents, administrative agents are bounded by limitations on their abilities to gather, store, and process information about their immediate decision-making context. They cannot, therefore, be maximizers. How, then, do boundedly rational agents make decisions? They are *satisficers.*

SATISFICING

Suppose an agent has categorized all outcomes of a decision problem as either "satisfactory" or not. It is then an easy matter to survey the options to see whether any option has a satisfactory outcome for every possible state of the world. Any such option is designated a satisfactory option. If there is exactly one such option, the boundedly rational agent chooses it. There is no guarantee, however, that any option will be satisfactory or that there is not more than one satisfactory option. And so the model of a satisficer is still incomplete.

To take the easy case first, what if there is more than one satisfactory option? Well, if each of several options is satisfactory, it makes no difference to the agent which it chooses. It might as well choose by any method whatsoever, such as taking the first one in the list. Or the one with the shortest description. To make the choice less arbitrary requires more input into the decision problem.

Suppose, then, that the agent is able to provide a strict rank ordering of all possible outcomes. The satisfactory outcomes can then be characterized as those at or above a given rank. This rank, by analogy with the psychological notion of an "aspiration level," is called the agent's *satisfaction level*. With this additional structure to the problem the simplest strategy for eliminating superfluous satisfactory options is for the agent to raise its satisfaction level until all but one option have been ruled unsatisfactory. A strict rank ordering of outcomes is sufficient to guarantee that a uniquely satisfactory option will be found. This simple strategy agrees with the psychological finding that humans tend to raise their aspirations whenever satisfaction of their current desires is easy.

What, then, if there are no satisfactory options? Here the agent has several possible courses of action. One is simply to lower its satisfaction level until some option becomes satisfactory. This strategy is guaranteed eventually to succeed so long as the agent can strictly rank order the outcomes. And it accords with the psychological finding that humans tend to lower their aspirations when satisfaction of their current desires is difficult.

But what if the agent does not wish to lower its satisfaction level? In this case it may proceed by seeking new options. By enlarging its set of options, the agent may discover a hitherto neglected option that is in fact satisfactory. This strategy is not guaranteed to succeed, however. If the agent is to make a choice, it may still have to lower its satisfaction level.

Finally, the agent can seek new information about possible connections among the options, states, and outcomes. Perhaps it can discover that some otherwise unsatisfactory outcomes will not occur, are physically impossible, or are at least highly improbable. This process may succeed in converting an otherwise unsatisfactory option into a satisfactory one.

Much more could be said in general about satisficing strategies, but further elaborations will be better appreciated in the more concrete context of choosing scientific models.

ARE HUMANS SATISFICERS?

As we have seen, there is extensive experimental evidence that humans, even scientifically trained humans, are not in general Bayesian agents. Is there similar evidence that humans are satisficers? Unfortunately not. This question simply has not been subjected to extensive experimental testing of the sort directed at the Bayesian model. Economists do not conduct this sort of research, and psychologists have not been sufficiently familiar with satisficing models to be concerned with their applicability. Of course, this also means there is no decisive evidence *against* the applicability of satisficing models, but that is small comfort.[4]

The main use of satisficing models, by Simon and his collaborators, has been in related research on human problem solving (Newell and Simon 1972). A good example of this sort of research is the study of chess playing by both novices and experts. Except very near the end of the game, the number of possible continuations in chess is so large that it precludes exploration of the whole tree structure even by the largest existing computers. Of course, real chess players do not proceed by searching the whole tree for the optimal move. Instead, according to Simon, they search through only a few (at most ten) moves. Their search tends to end not when they find the optimal move, but simply when they discover a *satisfactory* move. And they determine whether a particular move is satisfactory not by exhaustive study of that move, but by heuristic rules of thumb, aided by their memory of similar games already played. One of the main differences between novices and experts is simply that experts have a much greater store of previous games on which to draw.

Evidence that this is the right model of chess playing is of several sorts. One consists of interviews and "think aloud" protocols. Another is computer simulation. This latter method consists in constructing computer programs using satisficing principles and then implementing the program to see if it can simulate human playing. In both cases the evidence is encouraging, though probably not decisive.

We cannot, therefore, simply conclude that scientists are satisficers because people in general are satisficers. The case for scientists will have to be made on its own terms.

BOUNDED RATIONALITY AND NATURAL COGNITIVE ACTIVITIES

Because Simon is concerned with *rationality*, albeit bounded, it might be thought that his model is not appropriate for any "naturalistic" approach to the choice of models in science. But this is not at all the case. Simon used the word 'rationality' in framing his theory because he began with traditional economic models that define rational agents. His understanding of rationality, however, carries none of the ultimate, categorical connotations found in philosophical uses of the concept. Rational behavior, for Simon, is merely behavior directed toward the attainment of particular goals. The more effective a behavior is in achieving the agent's goals, the more rational it is. In short, Simon is concerned merely with "conditional," or "instrumental," rationality. Nothing is lost if we drop the term 'rationality', together with all its unwanted connotations, and simply talk about human actions directed toward reaching specified goals. This is just a type of natural cognitive activity.

There is, in fact, a strong parallel between Simon's project and the project of this chapter. Simon was led to his conception of satisficing because he found that the standard economic models of rational decision making simply did not apply to actual cases of decision making by administrators and other decision makers. I have been led to consider the virtues of satisficing at least partly because some prominent accounts of rational theory choice, which are very like the standard economic models of rationality, seem not to apply to the actual process of theory choice by real scientists. The question is how well Simon's model works in the scientific context.

Scientists as Satisficers

Now let us redeploy Simon's model, interpreting his agents as scientists, such as nuclear physicists, attempting to choose from among several models of a particular type of natural system, say nuclei with zero intrinsic spin. As before, we shall suppose that there are just two such models, a "standard" model, S, and a "new" model, D. I will presume a realistic interpretation of the possible states of the world. These are: (1) that the real systems in question are similar (in specified respects and degrees) to the standard model, or (2) that they are similar (in specified respects and degrees) to the new model. The options we shall interpret respectively as: (A) concluding that the standard model is most like the world, or (B) concluding that the new model is most like the world. The resulting two-by-two matrix is that already shown as figure 6.2.

VALUES AND INTERESTS

One immediate and very significant consequence of thinking about the choice of theoretical models in a decision theoretic framework is that the necessity for dealing with values or interests is explicit from the start. A decision theoretic model literally requires that there be a value rating attached to the outcomes of the decision matrix. There is thus no question *whether* values enter into the choice of scientific models, or even *how*— they must be attached to the outcomes of the decision matrix. The only question is *what* values are employed.

If we are talking about decisions by individual scientists, the values involved would be those of the scientist making the decision. But certainly not all values are idiosyncratic. Many are acquired through processes of acculturation and professionalization.

Looking at the outcomes in figure 6.2, one sees that there are two possible outcomes that correspond roughly to a "correct" decision. These are accepting the standard model if it fits the world and, alternatively, accepting the rival model if it fits. The other two outcomes represent "mistakes"—having accepted a model as fitting when in fact its rival is a better fit.

Philosophers who have conceived the acceptance of hypotheses in decision theoretic terms have introduced the notion of epistemic or scientific values (Hempel 1960; Levi 1967, 1980). They have reached little agreement, however, on just what are the epistemic values—except for truth. Scientists, it is said, value truth over error.

I prefer to speak of the similarity between a model and the world rather than the truth of statements, but the intent here is the same. And, indeed, my formulation of the decision problem makes it difficult to imagine how any scientist could prefer either of the mistakes to either of the correct outcomes. "My country right or wrong" is a statement of patriotic virtue. But "My model right or wrong" cannot be a statement of scientific virtue. No scientist could agree that one model was clearly a better representation of specified aspects of the world but nevertheless prefer to accept the other as being the better representation of those same aspects. Agreeing that science is a representational activity commits one to at least this much in the way of an evaluation of outcomes in the decision problem.

Unless we introduce some other values of some kind, we are left with a model of the disinterested seeker-after-truth encapsulated in the Mertonian norms. Such a scientist would value correct outcomes more highly than mistakes but would be indifferent to the choice between the two correct outcomes (and also, presumably, indifferent to the choice between the two mistakes).

Recent sociologists of science are generally agreed that the disinterested Mertonian scientist is a myth. But what does this mean in terms of our matrix? It means that actual scientists do not value the two correct decisions equally or regard the two mistakes with equal anxiety. That is, a scientist will normally prefer that the one model be accepted rather than the other and fear making one kind of mistake more than the other. This means that scientists must have some other interests in addition to their interest in being correct.

Philosophers have traditionally sought these additional values among such scientific virtues as simplicity, precision, and wide applicability. Even Kuhn (1977, chap. 13) has lately assigned these virtues an important, though not determining, role in the choice of theories. Recent sociologists look to professional and other broader interests. A scientist's expertise and professional standing may be wrapped up in one sort of model. Similarly, scientists may prefer one sort of model over another because it fits better with their more generalized social or metaphysical commitments.

One of the major general lessons from recent research on human judgment (Nisbett and Ross 1980) is that people are very poor at distinguishing among the various factors that influence their judgments. Since scientists obviously have both professional and social interests, any model of scientific decision making that restricted consideration to some supposed set of "scientific values" would stand little chance of fitting the actions of real scientists. Not that the presumed "scientific" virtues should be ignored. They simply must take their place alongside other sorts of values or interests.

Rejecting attempts to segregate supposed scientific values from other sorts of interests is very much in the spirit of satisficing models of decision making. Just as our model of scientific decision making cannot place unrealistic demands on the ability of scientists to gather and assimilate information, neither can it place unrealistic demands on the ability of scientists to distinguish among the influences of all the various values and interests that undoubtedly do influence even the "purest" of scientific decisions.

THE MINIMALLY OPEN-MINDED SCIENTIST

So far, then, our model of scientific decision making has the scientist valuing correct outcomes over mistakes, but otherwise free to place any value ranking or measure on the outcomes of the decision matrix. This freedom, however, is not consistent with satisficing strategies. Applying a satisficing strategy requires the additional restriction that the outcomes representing correct decisions be regarded as *satisfactory,* while those representing incorrect decisions are regarded as *unsatisfactory.*

This is not a trivial restriction. One can imagine a scientist whose career and skills are so bound up with a particular model that to reject the model would make it impossible to go on doing science. Such a scientist might well regard neither outcome associated with the acceptance of the rival model as being satisfactory. The only satisfactory outcome is accepting the favored model. This might describe the situation of famous "holdouts," like Priestly, who continued to accept their old model long after the vast majority of the profession had given it up.

Let us label these scientists not as irrational but merely as *closed-minded.* They simply will not countenance the rejection of their preferred model. By contrast, a (minimally) *open-minded* scientist will be one who at least regards both correct outcomes as "satisfactory." This does not

	S FITS BEST	D FITS BEST
CHOOSE S	SATISFACTORY	UNSATISFACTORY
CHOOSE D	UNSATISFACTORY	SATISFACTORY

Figure 6.5. A decision problem for an "open-minded" satisficer choosing between two models.

mean the scientist values both correct outcomes equally. The open-minded scientist need not be totally disinterested. If we assume a cardinal scale for the outcomes, the open-minded scientist might place a far greater relative value on one of the two correct outcomes. But the other outcome would still be regarded as "satisfactory." Figure 6.5 shows the resulting decision matrix for an open-minded satisficer.[5]

As it stands, the decision matrix of figure 6.5 has no satisfactory option. One outcome for each option is unsatisfactory. Lowering the satisfaction level is obviously no solution. But searching for additional options is possible. And, indeed, the situation might superficially be remedied by introducing the additional option of suspending judgment. If we suppose this option is satisfactory no matter which model best fits the actual world, we have a satisfactory option. The resulting model of scientific decision making might well capture some actual scientific contexts, but it obviously cannot be applied to all. Scientists cannot always suspend judgment, or else no theory would ever be chosen over any other. We need a model of scientific decision making that at least makes it possible for one of the substantive options to be chosen.

The remaining strategy for dealing with a decision problem lacking a satisfactory option is to search for new information. This is obviously a relevant strategy in a scientific context. The problem is to determine what kind of evidence is required and how it changes the decision problem so as to make the choice of one model a satisfactory option.

COGNITION AND INTERESTS

The above considerations show that there need be no conflict between a cognitive theory of science and the interest theories associated with the Edinburgh school. Indeed, a decision theoretic approach to understanding scientific judgment provides a way of giving structure and precision to explanations of scientific developments that appeal to "interests." And so what we have here may be viewed as a version of an interest theory.

The appearance of conflict arises because of doctrines that have been *associated* with interest theories, but are *not essential* to an interest theory as such. These doctrines are anti-realism and relativism. Yet it is not essential to an interest theory that there be no such thing as a representationally correct decision, or that scientists have no interest in being correct in a representational sense. What is essential is that scientists have other interests as well, and that these play a significant role scientific decisions. As I understand it, a cognitive theory of science need not deny the importance of these other interests. If it did, it could not be an adequate theory of science.

Experimental Tests

As Hacking (1983) has emphasized, experimentation has no single purpose. Some experiments are done merely to test new equipment. Others are performed simply to measure something that has not been measured before. But occasionally experiments are undertaken, at least in part, with the explicit purpose of *testing* rival models. When this happens, what is the connection between the result of the experiment and decisions about the relative fit of the rival models?

An experimental setup is a causal system. At least some parts of the system must be open to direct, physical manipulation by the experimenters. And the process must have a known range of possible results—a range that is constrained by the physical nature of the setup itself. It is a helpful abstraction to think of the parts of the process that can be manipulated as the *inputs* (or initial conditions) of the experiment, and of the results as being *outputs* (or final conditions).

The connection between the models under investigation and the output of the experiment is provided by the fact that the physical system being modeled must be a causally relevant part of the experimental setup. Moreover, for any model under investigation the experimenters must be able to determine what range of outputs would most likely result if that model in fact fits the system being modeled.

To make things more specific, let us continue with the supposition that there are only two models at issue: a standard model, S, and a "new" model, D. In addition, suppose that, given the specified experimental setup, a range, RS, of possible results is highly probable if the standard model (S) is roughly correct. Similarly, given the same experimental

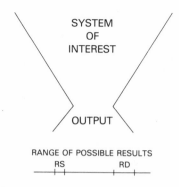

Figure 6.6. A schematic representation of an experimental test, with disjoint ranges of possible results for two different models of the system of interest.

setup, a range, *RD,* of possible results is highly probable if the new model (*D*) is roughly correct. Figure 6.6 pictures this setup schematically.

If we are to apply a satisficing strategy to the decision problem, there is one more set of conditions that must be realized in the experimental setup. If the standard model (*S*) provides the better fit, it must be highly improbable that the experiment will yield results in the range *RD*. Likewise, if the new model (*D*) provides the better fit, it must be highly improbable that the experiment will yield results in the range *RS*.

The only remaining relevant consideration is the obvious supplemental decision rule:

If *RS* is the actual result of the experiment, choose the standard model (*S*).
If *RD* is the actual result, choose the new model (*D*).

Now let us reconsider the decision problem pictured in figure 6.5 in the light of all these additions.

First, suppose the real world system is in fact roughly similar to the standard model. It is then highly probable that the experiment will yield result *RS*. Following the supplemental decision rule, the scientist will conclude, correctly, that the world is roughly like model *S*. Alternatively, suppose the real world system is in fact roughly like the new model. In this case it is highly probable that the experiment will yield result *RD*. Following the supplemental decision rule, the scientist will conclude, correctly, that the world is roughly like model *D*.

In either case it is highly probable that the scientist, a minimally openminded satisficer, will end up with one of the two satisfactory outcomes.

Correspondingly, ending up with either of the two unsatisfactory outcomes is highly improbable.

Here we go beyond the original Simon model only to the extent that we are now dealing with a high *probability* of achieving a satisfactory outcome rather than with the outcome itself. The values in the matrix are therefore expected values, the product of a utility and a probability. This means that both the original satisfaction levels and the model-based probabilities for getting results in the predicted range will have to be appropriately high. Otherwise, the expected value of an originally satisfactory outcome might no longer be satisfactory.

GIVING MURPHY HIS DUE

I am told by physicists that Murphy's law for experimentalists is that the data you get are always the least informative possible. If this is true, one would expect the data in our imagined experimental test to turn up in some region of possible results other than either *RS* or *RD*—say, right between these two regions. The supplemental decision rule stated above does not cover this case. My view is that if one does indeed get Murphy's result, the decision problem, as formulated, becomes inoperative. It is "back to the drawing board."

One could model this circumstance explicitly by introducing a third option, "Suspend judgment," and a corresponding clause in the decision rule: If the result is neither *RS* nor *RD,* suspend judgment. But this clause seems to me unnecessary and, in any case, unrepresentative of the situation as perceived by scientists. The possibility of an inconclusive result is always "in the background," as evidenced by the joke about Murphy's law. But "in the background" is the correct description. The third option comes to the fore only if one does indeed get Murphy's result. The supposition that one should get either *RS* or *RD* functions more like a *presupposition* of the experimental test than an explicit assumption. Its failure is treated accordingly.

In the end, this is not a point worth much argument. Anyone who wishes to keep things neat (I would say, unrealistically neat) is free to add the extra option and its corresponding decision rule.

MODEL-BASED PROBABILITY JUDGMENTS

The above model of scientific decision making employs two types of information not represented in the original decision problem depicted in figure 6.5. One is the result of some experiment. I should be the last to assert that determining what is in fact the result of any experiment is an easy matter. I shall, however, pass over that problem here because more serious problems face us. These concern the second type of new information, which is summarized in the required *probability judgments*. How are we to understand these judgments?

Having just been presented with much evidence that humans are not in general Bayesian agents, the reader might well wonder whether the presence of probability judgments in my satisficing model does not make it subject to the same objection. I think not.

One of the results of Kahneman and Tversky's investigations, exhibited, for example, in the taxicab experiment, is that humans are fairly good judges of the probability of events so long as they can interpret the events as the result of some causal process. That is, if people have some *causal model* they can use in making the requisite judgments, they are in fact good judges of the probability of events—relative to the model employed.

Now the probability judgments required by a satisficing model of scientific decision making are exactly of this sort. The scientist must judge the probabilities of various possible outcomes of some definite experiment, supposing that a given well-defined model is more or less correct. This is the only type of probability judgment required. In particular, the satisficing model does not assume that the scientist is capable of making a *diagnostic* inference, that is, using the experimental results to attach a probability to the model itself. It is just such diagnostic inferences that typify the Bayesian approach.

Of course, it is not enough to understand only the models under investigation. Since it is the probability of an experimental result that must be judged, the scientist must also have firmly in mind a precise model of the experimental setup. But this knowledge is generally available since experiments are, after all, designed, constructed, and carried out by the scientists themselves.

Philosophical Objections

A satisficing account of how scientists choose one model over another meets the goal of being both "naturalistic" and "realistic." It appeals only to natural decision strategies, and it permits hypotheses to be understood realistically.

At this point it is worth pausing to consider several philosophical objections, because they highlight differences between a satisficing approach to theory choice and more standard philosophical approaches.

THE ARGUMENT FROM ALTERNATIVE MODELS

Suppose, this objection goes, that in fact neither model S nor model D provides a very good fit to the system in question. Unbeknownst to anyone, a better fit is provided by a third model, T, which, as it happens, is overall more similar to S than to D. But in the given experimental situation, T predicts result RD, the same as D. Thus, when the experiment yields result RD, as it most probably would, one ends up accepting model D as the better fitting model, when in fact T, and even S, fit better. Moreover, and this is the clincher, unless either S or D already provides a *perfect* fit, it is a matter of logic that there will always exist (in the mathematical sense of 'exist') some such third model. The conclusion is that any decision to accept D as providing the better fit, on the basis of the experimental result RD, is *unjustified*.

This argument is somewhat softened by the realization that what would be accepted is at most the claim that model D is *similar* to the real world in specified respects and degrees. This might well be true in the case described, even though overall S fits better. But this reply fails to get at the heart of the objection.

What drives this sort of objection is the assumption that to justify any theoretical claim one would have to provide reasons for thinking that the accepted model is not only better than alternatives under active consideration, but also better than all *logically possible* alternatives, most of which will never even be thought of by anyone. But this assumption imposes a standard that is definitely not met in the actual practice of science. Its imposition leads to empiricism, or, worse yet, to more serious forms of skepticism.

Here it is important to keep in mind the lessons of the laboratory. When Rutherford adopted his model of the proton, he obviously did not have grounds for rejecting all possible alternative models. But the subsequent development of cyclotrons for use in nuclear research proved him right. Of course, it is logically possible that scientists in Rutherford's position were just lucky and not justified in adopting the models they did. And it is logically possible that cyclotrons work even though nothing like protons exists. But to point out these possibilities is once again to insist on a standard of justification that requires ruling out all logically possible alternatives.

Satisficers simply do not worry about such abstract possibilities of error. If someone were actually to develop a model like T, then one might try to design a different experiment to help decide between D and T. Lacking any explicit alternative models, one accepts the *possibility* that one has gone seriously wrong in spite of one's best efforts.

If it turns out that scientists are at least sometimes satisficers along the lines of the model just presented, it would then be an interesting fact about science as actually practiced that it sometimes succeeds in producing very well-fitting models in spite of the ever present abstract possibility of serious error. This fact would require a *scientific* explanation. No other kind of explanation would do.

A HUMEAN ARGUMENT

Among the recent arguments against realism is one that strongly resembles Hume's famous argument against the justifiability of induction (Cartwright 1983; Fine 1984a, 1986). Its seriousness for realists has been acknowledged even by prominent defenders of realism (Boyd 1984). The argument goes like this.

The justification of realistically interpreted hypotheses requires a principle of inference sanctioning a move from "success" to "truth." The hypothesis successfully predicts an experimental result, and one concludes (with appropriate qualifications) that it is true. Sometimes this is called "inference to the best explanation." But (said in a Humean tone of voice) what justifies this principle of inference? The almost automatic realist response is that the principle is justified by the success of science in using such a principle. But this response, the anti-realist replies, employs the very principle of inference whose justifiability is at issue. The realist response, therefore, begs the question. The challenge to the realist, then, is to produce some other response.

The proper realist response is to reject the challenge on the grounds that it is impossible to fulfill. A realist interpretation of science requires no such ultimate justification. A satisficing account shows how decisions regarding realistically interpreted hypotheses can be made. And it shows that this is an effective way of making such decisions because one has a high probability of reaching a correct conclusion. Furthermore, although one uses "success" to reach "truth," no "principle of inference" sanctioning a move from success to truth need be invoked. There is nothing left over that needs to be justified.

This is not to say, however, that a satisficing model provides a response to the Humean challenge that does not beg the question. The model-based probability judgments required by a satisficing strategy are themselves theoretical hypotheses. A satisficing account takes it as a scientific fact that such judgments can reliably be made. Any particular judgment can be questioned, of course, but only in the specific way that any empirical hypothesis might be questioned. There is in fact no serious threat here of infinite regress—only the Humean possibility of such regress.

SAFETY, STRENGTH, AND PROBABILITY

Another standard sort of argument has recently been employed by van Fraassen in support of constructive empiricism. Take any hypothesis that asserts a similarity between a model and a real system. A realistic version of this hypothesis, which claims similarity for theoretical as well as empirical aspects of the model, has more content than an empiricist version that claims similarity only for the empirical aspects. The realist version of the hypothesis is logically stronger than the empiricist version because it includes the empiricist version and more besides.

Now, it is a theorem of a properly interpreted probability calculus that for any fixed body of evidence, the conditional probability of the weaker, included hypothesis must be greater than that of the stronger, including hypothesis. In short, there is an inverse relationship between logical strength and probability. If one interprets the conditional probability of a hypothesis as measuring something like its "degree of evidential support," it follows that, no matter what the evidence, the empiricist version of any hypothesis is necessarily better supported than the corresponding realist version. That, it is concluded, is a reason always to prefer the empiricist version.[6]

Strictly speaking, a satisficing account of how scientists choose hypotheses bypasses this argument. This account employs no conditional probabilities of *hypotheses*. It does not even presuppose that such probabilities are well defined. A satisficing account employs only model-based probabilities, which at most might be understood as conditional probabilities of the *data*, given various hypotheses. Within a satisficing framework, therefore, one is not choosing the less well-supported hypothesis because this measure of support is not present in the framework.

Yet the spirit of the objection can be formulated in more general terms. It is surely true that the empiricist version of a theoretical hypothesis is "safer" than the realist version in the sense that it is less likely to be proven false. Since it claims less, it carries with it fewer ways in which it could be false.

All that follows, however, is that there is a price to be paid for preferring realistically interpreted hypotheses. One is more likely to be proven wrong in the end. It does not follow that scientists are not generally willing to pay that price, or that they should not do so. Nor does it follow that the benefits of having realistically understood hypotheses are not worth the price.

I suspect that the reason the argument under discussion appeals to empiricists is that they believe no benefits accrue to realistically interpreted hypotheses. The supposed benefits, they think, are pure illusion. A realist, to the contrary, would argue, for example, that one of the advantages of understanding Rutherford's proton model realistically is that it led others to think about how to design a machine—a cyclotron—that could be used to bombard nuclei with protons. That one could understand this development adequately in an empiricist framework is, the realist would say, no more than empiricist science fiction.[7]

The Role of Probability in Science

One of the central problems in the logical empiricist tradition concerned the proper interpretation of probability (Salmon 1967). The reason this problem seemed so important is obvious. Probability was seen as the key to scientific inference, and thus as a crucial part of the methodological basis for the *justification* of scientific claims. Echoes of this earlier program, particularly its Carnapian (1950) version, still resound in recent philosophical writings. Probability continues to be thought of as "the logic of rational belief," not only by Bayesians (Jeffrey 1985), but even by philosophical critics of strict Bayesian accounts of scientific inference (Kyburg 1961, 1974).

My view is that probabilistic models are not fundamentally different from any other scientific models, the models of classical mechanics, for example. Their main distinction is that they are applicable in a wide variety of different contexts ranging from physics to sociology. Partly this is because of their extremely simple structure. But it is also because that simple structure appears frequently in nature itself.

In support of my position I should like to present briefly some research on judgments concerning simple mechanical systems. This research on "the intuitive physicist" exactly parallels that on "the intuitive statistician," though it is not nearly so extensive. In fact, similar research on "the intuitive logician" suggests that even much of what we are pleased to call logic is best thought of as a family of models that must be learned and whose applicability to the world is a contingent matter. In this view not even logic is "like logic." Rather, logic is "like probability," which, in turn, is "like physics."

The Intuitive Statistician Revisited

The attitude that probability should function as a kind of logic shows up even in the empirical research on human judgment. Probability theory, Bayes's theorem in particular, is regarded as providing the normatively correct standard against which actual human reasoning is to be measured. When human reasoners fails to reach the normatively correct conclusions, it is hypothesized they are applying various "heuristics" that, while fairly useful in many ordinary situations, introduce "biases" that in some circumstances lead to grossly mistaken judgments. In any case, the failures produced by such biases are regarded as failures in *reasoning* or judgment. The difference between the intuitive statistician and the trained statistician or social scientist thus tends to be regarded as a difference in reasoning ability—perhaps even a difference in rationality (Nisbett and Ross 1980; Kahneman, Slovic, and Tversky 1982).

Cognitive psychology itself contains the resources for a somewhat different perspective. It is now generally agreed that people approach any situation with a store of models ("schemata"). And although the capacity for employing models, perhaps even general types of models, is biologically based, the content of the particular models available to any individual is acquired from the surrounding culture. How well people perform in any given context thus depends on the adequacy of the models at their disposal, as well as on their skill in deploying those models.

This difference in perspective is nicely illustrated by the taxicab problem. Kahneman and Tversky suggested that people are naturally inclined to employ *causal* models and are less good at "diagnostic" reasoning. Indeed, people often make serious mistakes in judgment because they employ an incorrect causal model while ignoring relevant diagnostic information. The tendency to favor causal reasoning is thus seen as a biasing factor in human judgment.

I have a slightly different explanation of why people ignore the base rate. It is that most people simply do not have at their disposal any models that make it possible for them to process base-rate information. They simply do not know what to do with it. And so, not being able to process the base-rate information, they ignore it and use instead the information they can process, namely, in the cab case, the reliability of the witness.

This interpretation is supported by the causal variant of the cab problem in which the subject may easily invoke the idea that the Green cabs are mechanically defective or that their drivers are reckless. People certainly do possess a variety of such causal schemata, and they deploy them quite effortlessly. The causal schemata provide a vehicle for using the base-rate information, if only imperfectly. And in fact it is used.

By paying more attention to the models actually available to subjects, we gain a better picture of the difference between intuitive statisticians and trained social scientists. The latter have at their disposal a whole family of probabilistic models they can deploy in handling the sorts of problems presented. And, as a matter of fact, rather than of logic, these models are better fitted to the kinds of cases at issue.

Being able to deploy probabilistic models is not always an easy matter, however, which is why even trained statisticians sometimes make mistakes similar to those of intuitive statisticians. A final example illustrates this point. Kahneman and Tversky (Kahneman, Slovic, and Tversky 1982, 495) reported having asked "many" squash players the following question:

> As you know, a game of squash can be played either to 9 or to 15 points. Holding all other rules of the game constant, if *A* is a better player than *B*, which scoring system will give *A* a better chance of winning?

All the respondents were said to have "some knowledge of statistics." "Most" said the scoring system should not matter.

A squash game, however, can be considered as a series of binomial trials with the result that either *A* or *B* wins the point. If on each trial *A*'s chances of winning the point are somewhat greater than *B*'s, the probability of *A*'s being ahead *increases* with the number of trials. In statistical terms the variance decreases with increasing sample size. Thus, *A* should prefer the 15-point game.

The researchers reported that when presented with this argument almost all of the respondents immediately acknowledged their initial response had been mistaken. The authors took this response as evidence that the initial mistake was one of application rather than comprehension.

I would suggest that part of "comprehension" is a matter of having the relevant models available in long-term memory. "Application" is a matter of being able to retrieve and deploy the appropriate models when needed. When asked the question, the respondents tended to call up a batch of squash-playing schemata. None of those schemata suggested a definite answer to the question, and so the respondents were inclined to say that it did not matter. They tended not to call up statistical schemata. Nothing explicit in the question suggested that such schemata might be relevant. Moreover, if the context in which the question was asked were a squash court or cocktail party rather than an academic office, the general failure to employ statistical models would be even more likely. In any case, being presented with the statistical analysis called forth the appropriate schemata, whose applicability the respondents immediately appreciated.

There is no mention of this problem's being tried on statistically *naive* subjects. It seems very likely, however, that their initial response would have been similar but that they would not immediately have appreciated the statistical analysis that followed.

In sum, my claim is that naive subjects who make mistakes in simple statistical problems are not lacking in *reasoning* ability, but simply in *knowledge* of the most appropriate models for the situation.

THE INTUITIVE PHYSICIST

Several investigators have recently conducted experiments on college students to determine their conception of simple mechanical phenomena of the type studied in elementary physics courses. These researchers have given no hint that they think of themselves as investigating students' reasoning ability. They conceive their aim as the discovery of the models people actually employ in dealing with simple mechanical systems. Yet there is a "normatively correct" model in the background. It is that of Newtonian mechanics. And, as one might expect, significant divergences from the Newtonian model can be demonstrated in very simple experiments.

One of the favored strategies in this research is to present the subject with a drawing of a dynamical system poised at a well-chosen instant and to ask the student to draw the subsequent motion of the system. Figure 6.7 shows two such drawings. The instructions applicable to these drawings were (McCloskey 1983, 300):

> The diagram shows a thin curved metal tube. In the diagram you are looking down on the tube. In other words, the tube is lying flat. A metal ball is put into the end of the tube indicated by the arrow and is shot out of the other end of the tube at high speed.

The general instructions were to draw the path of the ball as it emerges from the other end of the tube.

In one experiment 50 students were presented with four diagrams representing similar situations. Of the 47 who completed the exercise, 15 had taken no high school physics course, 22 had completed one high school physics course, and 10 had completed at least one college physics course.

The correct response, of course, is that the emerging ball follows a straight line as it exits the tube. Yet 34 percent of all the respondents' diagrams showed the ball following a *curved* path as it exited the tube. Of drawings by students with no previous physics instruction, 49 percent showed curved paths. The percentages for the other two groups were 34

Figure 6.7. Two drawings employed in research on students' use of models in thinking about mechanical systems. Adapted from McCloskey (1983).

percent and 14 percent, respectively (McCloskey, Caramazza, and Green 1980, 1139).

Of particular interest in this research is that the "mistakes" are systematic. Of the 25 subjects who drew curved paths for either of the two drawings in figure 6.7, 19 showed a greater curvature for the spiral tube than for the C-shaped tube. The hypothesis considered is that those students had in mind a model of "circular inertia" similar to the impetus models developed by medieval scholars. The more curved the motion the ball has experienced, the greater its circular impetus.

We need not pursue these speculations here. The relevant point is that no one would be tempted to treat these mistakes as failures of reasoning or as being due to faulty "heuristics." They seem instead a matter of applying an inadequate model whose features are to some extent revealed by the experiments.

A similar attitude toward probability judgments might consider the possibility of a "representativeness model" in which probabilities are a function of representativeness. One could then investigate the details of this model and how it differs from standard probability models. There would then be no temptation to speak of failures of "reasoning" or "rationality." There would only be a lack of knowledge of the appropriate model, or a failure to apply it.

THE INTUITIVE LOGICIAN

The idea that mathematical or logical reasoning provides the paradigm of rationality goes back at least as far as Plato. And it was, of course, the ideal of a logical system that provided a main inspiration for logical empiricism. Challenging this ideal is not something to be undertaken lightly. Here I will only gently suggest that logic and mathematics, like mechanics and probability, provide a set of models whose applicability is contingent and whose use requires learning and practice.

I will present just one experiment of many, P. C. Wason's famous "selection problem." The subjects, British undergraduate students majoring in science and mathematics, were presented with cards like the rectangles shown in figure 6.8, where the hatched areas indicate the part of the card that is hidden. The instructions were as follows (Wason 1977, 119–20):

> Which of the hidden parts of these cards do you need to see in order to answer the following question decisively? FOR THESE CARDS, IS IT TRUE THAT IF THERE IS A CIRCLE ON THE LEFT, THERE IS A CIRCLE ON THE RIGHT? You have only one opportunity to make this decision; you must not assume that you can inspect

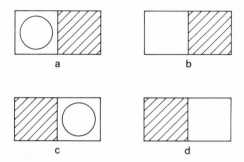

Figure 6.8. Wason's selection task. Adapted from Wason (1977).

cards one at a time. Name those cards which it is absolutely necessary to see.

This experiment, in many variations, has been tried on many subjects, with similar results. The "correct" answer is, of course, cards *a* and *d*. Either the failure to find a circle on right side of *a* or finding a circle on the left side of *d* would falsify the test statement. Whatever is on their covered parts, *b* and *c* are compatible with both the truth and the falsity of the test statement.

In one experiment subjects were presented individually with a version of the problem stated above. If they gave the wrong answer, the cards were uncovered and the correct answer revealed. If they still did not acknowledge the correctness of the answer presented to them, there followed an unstructured interview that proceeded until either the subject acknowledged the correctness of the revealed answer or the interviewer thought further discussion would be pointless. Of the 36 subjects in this experiment, only 2 gave the correct answer initially; another 9 acknowledged it upon being shown the uncovered cards: another 12 acknowledged it in the course of the interview; and 13 never acknowledged it at all. (Wason 1977, 122–23)

To anyone who has ever attempted to initiate college students into the mysteries of material implication, these results are hardly surprising. I have noted them only because of the parallel with similar experiments concerning probability and mechanics. The parallel suggests that we treat failures in such exercises not as evidence of deficiencies in reasoning ability, but simply as indicators of ignorance of the most appropriate models for the situation. For real people in real life, the models of logic seem not fundamentally different from those of physics. If this is true of logic, it is certainly true of probability.

A Computer Analogy

In the design and operation of computers one distinguishes between the operating system and working programs. Both may be software, of course, but the operating system is more fundamental. It provides the link between the hardware and all other programs. If one were to design an ideal robot scientist along traditional philosophical lines, one would make logic and probability part of the operating system. Physics would be a working program. If one were to design a robot more like actual humans, one would use a more primitive operating system. Logic, probability, and physics would all be working programs to be added later.[8]

In humans the operating system is not just software. At least part of it must be "hard wired," the product of evolution. Now, it may be that *fragments* of logic, geometry, and physics have likewise been built in by evolution. If that is true, it will have to be an *empirical* finding, not the result of a priori philosophical analysis.

7

Models and Experiments

Having developed a satisficing model of scientific judgment, I must now consider whether this model has any interesting applications. In accord with my own theory of science, it is not necessary to show that every scientific decision made by every scientist, even just those decisions involving the choice of models, is an instance of satisficing. It is enough that satisficing models fit some important cases—that sometimes real scientists are satisficers.

There is a prior, and even more difficult, question to be faced. How could one best go about determining whether a satisficing model fits actual instances of scientific decision making? One could, perhaps, devise experiments along the lines of those performed by Kahneman, Slovic, and Tversky (1982), and then try them on working scientists. Similar experiments based on Wason's (1966) selection task have been done with working scientists (Tweney and Yachanin 1985; Griggs and Ransdell 1986). Or one could follow Simon and his associates (Newell and Simon 1972; Ericsson and Simon 1984) in presenting fabricated scenarios to scientists and then performing a protocol analysis on their responses. One *could,* of course, do all those things, and more. And the results might be very informative. But I do not possess the resources to carry out such investigations.

I shall begin, therefore, with the traditional method of the historian or philosopher of science, the analysis of scientists' writings. In addition, having had access over an extended period to scientists engaged in ongoing research, I can supplement the examination of both published and unpublished writings with more ethnographic sorts of data, particularly observations and interviews (in some cases repeated interviews with the same individual).[1]

A drawback to such an approach, of course, is its narrow focus. I shall be examining only a small part of one science. But this is not as limiting as it might at first seem. Although I shall emphasize experiments I have my-self observed, I shall not confine myself to those experiments. We shall be examining an extensive ongoing effort in current nuclear physics involving a number of researchers at various locations. Moreover, nuclear physics is itself a paradigm of twentieth century science. In its use of mathematical techniques, of computers and other advanced technology, as well as in its organization into research groups, nuclear physics resembles many other contemporary sciences.

Since I cannot presume that the reader knows anything about nuclear physics, I shall try to present the required theoretical background in a form understandable to the uninitiated. This is not an easy task, however, be-cause the relevant models derive mainly from quantum theory. When in doubt I shall oversimplify the physics in pursuit of general intelligibility.

Models of the Nuclear Potential

To gain some understanding of the physics behind the case we are about to discuss, it is helpful to recall Rutherford's original experiments at the very dawn of nuclear physics (Wilson 1983). Rutherford discovered the nucleus by directing naturally produced alpha rays at a thin metal foil and noticing, quite to his surprise, that some of the alpha particles were deflected back in the general direction of the source. He concluded that the target material must contain positively charged particles that were fairly massive relative to alpha particles—in short, nuclei. The alpha particle is repelled by the well-known Coulomb force, as pictured in figure 7.1. Like charges repel one another following an inverse square law of force.

Since the nineteenth century, physicists have tended to represent forces by potential fields and to talk about *potentials* rather than forces. The mathematics of potential fields are much easier to handle, and the forces are easily recoverable. For our purposes the only relevant fact is that posi-tive potentials correspond to repulsive forces and negative potentials corre-spond to attractive forces. Thus, Rutherford's alpha particles could be de-scribed as having interacted with the positive Coulomb potential of the nucleus.

Although the similarity between Rutherford's original experiments and contemporary cyclotron research is obvious, nuclear physics has come a long way in 75 years. The nucleus is now thought to include other particles in addition to protons, such as neutrons, a family a mesons, and a spectrum of quarks. Mesons and quarks are related to the strong nuclear forces that hold the nucleus together. The existence of such forces, though not their makeup, was obvious even as early as Rutherford, since without them the

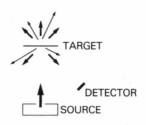

Figure 7.1. A schematic representation of Rutherford's experiment.

nucleus would not hold together. The positive charges would simply repel one another. Current models of the nucleus also exhibit considerable internal structure. The nucleons (protons and neutrons) are arranged in "shells," analogous to the orbits of electrons in the familiar Bohr model of the atom. Finally, since Rutherford there have been huge developments in the theoretical concepts used to model the *dynamics* of nuclear interactions, namely, quantum theory, quantum electrodynamics, and so forth. Fortunately, one need know little about the details of those theories to understand the case at hand.

THE NUCLEAR OPTICAL POTENTIAL

Our present case is like Rutherford's in that it involves the elastic collision of a positively charged particle (a proton) with a nucleus. In an elastic collision the projectile interacts with the target, but no other particles are emitted. The projectile itself (rather than some other particle) is detected after the interaction. The (*p, n*) reactions discussed in chapter 5 are, by contrast, inelastic.

The *nuclear optical potential* is the aggregate of the nuclear potentials of all the individual particles in the nucleus. Neglecting the Coulomb potential, what a proton passing by a nucleus experiences is the nuclear optical potential of that nucleus.

One might think that it is possible, at least in principle, to determine the optical potential of a particular type of nucleus simply by combining the potentials of all the individual nucleons. But this is not possible, not even in principle, because all the nucleons interact with one another, thereby creating a many-body problem. Even in Newtonian physics it is not possible to derive, in closed form, a complete dynamical solution to the problem of three bodies subject only to gravitational forces. The best one can do is write down the form of the terms for an infinite series. Any derivation of the optical potential from individual potentials must therefore be an *ap-*

proximation obtained by truncating the infinite series in some way or other. As a result all the fundamental models of nuclear physics incorporate explicit approximations. There are no fully specified models.

In general one "solves" a many-body problem by ignoring whole classes of possible interactions. Thus, for any type of nucleus there is a whole family of models of its optical potential indexed by which possible interactions are included (or excluded) in the calculation. The simplest such model, of course, is the one that ignores all interactions among nucleons in the nucleus. That approximation, which basically treats the nucleus as a set of "quasi-free" nucleons, is called the *impulse approximation*.

NONRELATIVISTIC SCHROEDINGER MODELS

For several decades, up to around 1980, the standard optical model was an impulse approximation based on the Schroedinger equation for two free nucleons. The Schrodinger equation is a quantum mechanical analogue of Newton's second law of motion. It describes dynamic interactions among atomic particles.

The Schroedinger equation is fundamentally nonrelativistic. It does not satisfy the spatial and temporal invariance conditions imposed by the special theory of relativity. Moreover, it does not allow for the creation or annihilation of particles following Einstein's famous mass-energy relationship, $E = mc^2$.

But surely physicists must believe that protons and nuclei obey the laws of relativity theory. Moreover, one might recall, even just 200 MeV protons move at half the speed of light, and that is fast enough for relativistic relationships to be important. How could the standard model be nonrelativistic? The answer has two parts.

First, there is a further matter of *approximation*, of ignoring something regarded as too small to measure. The kind of relativistic effects that would matter would be those involving the creation of particles in the nucleus. Now, the amount of energy it takes to create a nucleon approaches 1,000 MeV. But nucleons can be created only in pairs, requiring, therefore, a minimum of 2,000 MeV. To make matters worse, when a particle strikes a stationary target, half of its energy goes into the forward motion of the particle, leaving only half of its energy available for particle creation. Thus, a 400 MeV proton would have only 10 percent of the minimum energy needed for relativistic interactions. For detectable effects much more than the minimum would be needed. Of course, these calculations assume standard Schroedinger nuclear potentials.

Second, relativistic considerations are not totally ignored even though the basic form of the Schroedinger equation is nonrelativistic. The *kinematics* of relativity theory, including the fact that the incident protons are moving at relativistic velocities, is routinely employed.

But the inclusion of "relativistic" components in Schroedinger models goes even deeper. About 1930 Dirac developed a relativistic version of the Schroedinger equation, now known, of course, as the Dirac equation. This equation associates a magnetic moment with some free particles, the kind of thing one would get classically with a charge on the surface of a spinning sphere—whence the suggestive name 'spin'. Even in classical electrodynamics such a magnetic moment would interact with an electric current, such as that produced by a moving charged particle. Such interaction produces a "spin-orbit potential," which in some cases may be as strong as the central nuclear potential itself. Thus, even though their ultimate theoretical explanation derives from relativistic models, terms representing spin-orbit potentials typically appear in the nonrelativistic Schroedinger equation for nucleon-nucleon interactions.

Figure 7.2 shows a typical Schroedinger equation for a proton in the optical potential of a nucleus of calcium 40. Below the equation are graphical representations of the two potentials included in the equation. Both poten-

$$\left[-\frac{\vec{\nabla}^2}{2m} + V_{CENT} + V_{S.O.}\vec{\sigma}\cdot\vec{L} \right] \psi = E\psi$$

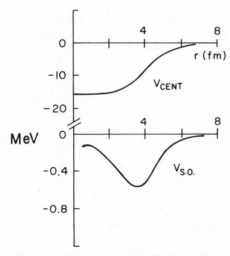

Figure 7.2. A Schroedinger equation and potentials for the elastic scattering of 200 MeV protons by calcium 40.

tials are negative, and thus attractive, with the spin-orbit component in this case being roughly one-tenth that of the central potential. The total optical potential in the model is the sum of those two potentials.

THE SCHROEDINGER APPROACH

The situation in nuclear physics nicely illustrates my earlier characterization of a scientific theory as being a family of models. There is no single "Schroedinger model" of the nucleus. Rather, there is a family of models whose members are all characterized by the general form of the Schroedinger equation, but differ in the types, and the details, of the included interaction potentials.

Physicists themselves have a standard way of referring to this broad family of models. They call it "the Schroedinger approach" or, sometimes "the nonrelativistic approach." When they wish to refer to smaller families of models or, indeed, to individual models, they use a brief characterization or the name of a person, or persons. Typically, these are coded in capital letters, like government agencies. Thus there is the "DWIA" (distorted wave impulse approximation), the "KMT" model (named for its authors, A. K. Kerman, H. McManus, and R. M. Thaler [1959]); "BHF" (Brueckner-Hartree-Fock) theory, and so on.

A philosopher-logician of the Stegmueller (1979) school might maintain that there must, in principle, be a super Schroedinger model that includes every known nuclear interaction potential. Every member of the Schroedinger family of models would then be a special case of the one super model. One could even imagine such a philosopher of science attempting to characterize this super model by means of a suitably framed set of axioms. Not only would such a reconstruction fail to capture nuclear theory as employed by physicists, it cannot even in principle be carried out.

The strongest reason it cannot is that already noted. Every attempt to model the optical potential must resolve the many-body problem by making some sort of approximation. Different ways of doing the approximation yield somewhat different potentials, many of which are not experimentally distinguishable. One cannot single out a preferred model in a nonarbitrary way. There is, nevertheless, a kind of unity to nuclear theory, but that unity is more organic than deductive.

Background to the Pursuit of Relativistic Dirac Models

No nuclear physicist seems to doubt that, in principle, the correct model of the nucleus would be a relativistic model based on the Dirac equation. The issue is whether interactions at energies typical of research in nuclear physics involve fundamental relativistic processes to an extent great enough to detect experimentally. If not, the research can proceed happily using non-relativistic models.

As noted above, the prevailing attitude among nuclear physicists before 1980 was that fundamental relativistic models were not necessary. Very little research, whether theoretical or experimental, focused on such models. Five years later, however, there was considerable interest in relativistic models. Whole conferences were organized to discuss the latest theoretical and experimental findings relevant to the development of these models (Shepard, Cheung, and Boudrie 1985). What produced this new interest in the pursuit of relativistic models?

From a cognitive point of view the question is: How did individual scientists make the decision to devote their energies to the development or testing of relativistic Dirac models of nuclear interactions? The answer is not quite the same for theoreticians as it is for experimentalists. In both cases, however, one can detect elements of a satisficing strategy.

A crucial factor for both theoreticians and experimentalists seems to have been the appearance of new data that failed to fit the prevailing Schroedinger models, but which were very quickly shown to fit preliminary Dirac models. Moreover, the new data were being produced by new instruments (which were originally designed for other purposes). In short, a decision theoretic analysis presents the new focus on Dirac models as originally instrument- and data-driven, though other interests were also crucial, as we shall see.

THE NEW DATA

In the mid- to late 1970s, several research groups were using elastic proton-nucleus (pA) interactions to study the density distribution of neutrons within the nucleus. One of the centers of this work was the Los Alamos Meson Physics Facility (LAMPF), which has a medium-high-energy proton accelerator (800 MeV). Because of the nature of the proton-neutron interaction, it was thought that interactions using spin-polarized protons might be particularly revealing. Thus, in 1977, several research groups working at LAMPF, particularly a group from the University of Texas (Hoffmann 1985), began collecting data using a newly constructed polarized ion source together with a new high-resolution spectrometer (for measuring the energy of scattered protons).

Before the development of polarized ion sources the basic measured quantity in elastic scattering experiments was the cross-section (really the "differential" cross-section) $d\sigma/d\Omega$. Roughly speaking, the cross-section at the scattering angle θ is simply the number of protons scattered between angles θ and θ plus a small increment $d\theta$, divided by the total number of incident protons. Experimentally, researchers measure the relative number of protons recorded by a counter located at an angle θ from the path of the incident proton beam. The standard experimental arrangement is pictured in figure 7.3.

Once one has a reliable source of polarized protons, one can measure a more complex parameter called variously polarization, *P*, or analyzing power, *Ay*. Figure 7.4 depicts the experimental setup for measuring polarization. The target and counters are arranged as in standard measurements of the cross-section. The difference is in the incoming proton beam. The spin of the protons is regulated so that the axis of rotation is perpendicular to the plane determined by the beam, target, and counter. The direction of rotation is alternated 180 degrees between "spin up" (represented by the dot in a circle—tip of the arrow) and "spin down" (represented by the cross in a circle—tail of the arrow). The polarization is a function of the difference between the cross-section measured with the incoming protons in the spin-up position and the cross-section measured with the incoming protons in the spin-down position.

The first polarization data, for 800 MeV protons on a calcium 40 target, were published in 1979 (Ray 1979). As shown in figure 7.5, those data

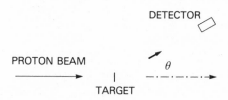

Figure 7.3. A schematic representation of the experimental setup for measuring differential cross-sections.

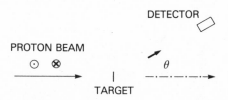

Figure 7.4. A schematic representation of the experimental setup for measuring polarization (analyzing power).

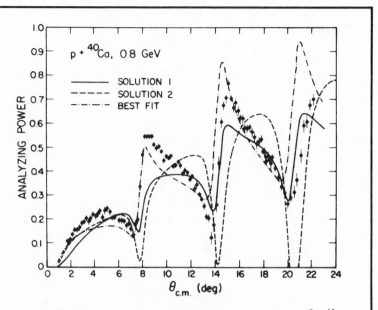

FIG. 5. Predicted elastic analyzing powers for $\vec{p}+{}^{40}$Ca
at 0.8 GeV compared to the data given in Ref. 4. The
prediction assuming solution 1 (2) for both the $p+p$ and
$p+n$ amplitudes is indicated by the solid (dashed) curve.
The dash-dot curve is the best fit obtained by allowing
a free variation in the $\bar{\sigma}_p$ and $\bar{\alpha}_{sp}$ parameters as discussed in the text.

Figure 7.5. The first data on polarization (analyzing power) measured at 800 MeV for a calcium 40 target. Reproduced from Ray (1979, 1860).

failed to agree with any standard Schroedinger model. The "best fitting"
curve (dot–dash), calculated directly from the data using standard least
squares techniques, obviously failed to fit calculations of P for two different Schroedinger models (the solid and dashed curves).

As a result of these and similar data, research at LAMPF shifted from
neutron density studies to an investigation of the Schroedinger models on
which the neutron density studies depended. Two consequences of that
reorientation were a move to a somewhat lower beam energy, around
500 MeV, and the design of a new polarimeter to accompany the high-
resolution spectrometer. Both developments were expected to provide a

good chance of revealing the source of the disagreement between the standard models and the 800 MeV data. At that point there seems to have been no thought of actually testing Schroedinger models against any radically different alternatives.

The polarization data at 500 MeV, for calcium 40 and several other targets, appeared in 1981. Those data, however, exhibited the same sort of divergence from standard calculations as did the data at 800 MeV. Switching to the lower energy did not help. The paper reporting the data (Hoffmann et al. 1981) was titled "Elastic Scattering of 500-MeV Polarized Protons . . . and Breakdown of the Impulse Approximation at Small Momentum Transfer." Obviously, these physicists thought that the problem lay in the excessive simplicity of the impulse approximation. They suggested examining various "nuclear medium corrections to the IA"—that is, introducing some interactions among nucleons into the basic Schroedinger model of the optical potential. But still there was no mention of any radical departures from the Schroedinger approach.

The new polarimeter made it possible to measure a second spin variable, the spin rotation parameter, Q. Figure 7.6 shows the layout for this measurement. Here the spin axes of the incoming protons are in the plane of, not perpendicular to, the beam, target, and detector. As in measurements of the analyzing power, the orientation of the spin vector of incoming protons is measured before their interaction with the target. The big difference in spin rotation experiments is that the spin of the interacting protons is measured again *after* the interaction—that is the job of the new polarimeter. The quantity Q is a function of the angle, β, by which the spin of the interacting proton is rotated as a result of the interaction. It is also a function of both the cross-section and the analyzing power. Thus, to measure Q for any target material requires making several sets of measurements for each angle.

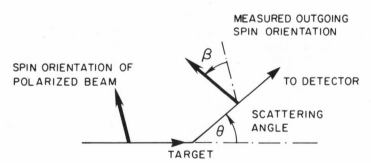

Figure 7.6. A schematic representation of the experimental setup for measuring spin rotation.

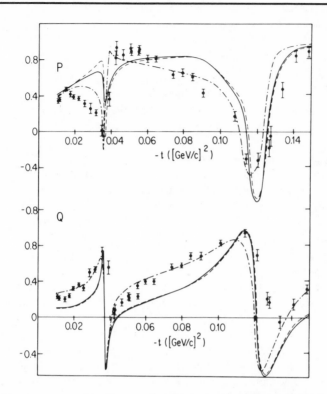

FIG. 2. P and Q for elastic scattering of 497 MeV
protons from ^{40}Ca. The solid line corresponds to the
full calculation done according to the approach of
Ref. 11 with the N-N amplitudes from the phase-shift
analysis (Ref. 12). The dashed curve was obtained by
neglecting the magnetic moment interaction. The dash-
dotted line correspond to the calculation with the Gaus-
sian spin-orbit amplitude. See Eq. (7) in the text.

Figure 7.7. The first data on spin rotation measured at 500 MeV for a calcium 40 target.
Reproduced from Rahbar et al. (1981, 1813).

The first published measurements of Q for calcium 40 at 500 MeV again
showed considerable deviation from standard predictions, as shown in figure 7.7. In referring to those data, the authors remarked that "our theoretical predictions for Q and P reflect a serious lack of agreement *quantita-
tively* with these 500 MeV data" (Rahbar et al. 1981). Again their sug-

gestions for what might be wrong were solidly within the Schroedinger approach.

HOW NUCLEAR PHYSICISTS JUDGE GOODNESS OF FIT

To this point researchers on the subject seemed agreed that there was a "serious lack of agreement" between the new data on the spin variables, P and Q, and the standard Schroedinger-based impulse approximation of the optical potential. Before going on it is worth pausing briefly to ask how they reached that judgment.

The answer—on the surface at least—is deceptively simple. Theoretical values of the desired parameters, in this case P and Q, are calculated using one or more standard models. The data points are plotted on the same graph with "error bars" representing the expected statistical variation in the data. In practice the expected statistical error is taken to be $\sqrt{n}$, where n is the number of events counted. One then *visually compares* the theoretical curves with the data points and judges whether the fit is "extraordinarily good," "very good," "good," "reasonably good," "lacking in agreement," "seriously lacking in agreement," and so on. One might also remark whether the data are in "qualitative" agreement with the predicted values, that is, whether they exhibit the same general shape as the predicted curve. Given qualitative agreement, further divergences are often referred to as being only "quantitative." Lack of qualitative agreement is regarded as much more serious. Basically, however, that is all there is to it.

This seems more like wine tasting than hard science! What about all those "goodness of fit" tests used by social scientists? Nuclear physicists rarely use them to judge goodness of fit. Publications in nuclear physics rarely contain values for χ^2 or any other such parameter, as anyone can verify by leafing through the nuclear physics section of the *Physical Review*. Nor, judging from my experience, do physicists use those measures informally. Such things simply play little role in nuclear physics.

The best explanation of this situation I have encountered was told to me by an experimentalist. Good experiments, he said, have a margin of error of around 2 percent. But hardly any model in nuclear physics comes within 20 percent of the data. All data in the field would therefore yield very large values for any standard measure of fit, and comparisons would not be very meaningful. "More kinds of data can be assimilated by the eye and brain in the form of a graph," he said, "than can be captured with χ^2." I immediately thought of geometrical figures being processed in the hippocampus (or wherever). Surely, this topic deserves further study.

Response to the New Data

If one were to employ a Kuhnian account of scientific development at the level of specialties and research groups, one would describe the above development as a case of normal science interrupted by anomalies and crisis. One would then predict a period of revolutionary science, during which scientists generate new ideas, followed by a period of competition among rival approaches, leading eventually to a renewed period of normal science based on a new "paradigm." But that is not the pattern found here. The Lakatosian or Laudanian pattern of continuing, competing research traditions is somewhat closer to the mark.

Closer still is an evolutionary, or ecological, account in which variant theoretical approaches coexist in stable equilibrium, with one approach clearly dominant. The variant approaches become active rivals only when changes in the scientific environment alter their relative fitness. What any such evolutionary account requires is a mechanism by which the ensuing competition takes place. This mechanism is individual decisions.

The basic elements of relativistic Dirac models of the nucleus had existed for 50 years. Moreover, these elements were fairly well known and had been elaborated by various individuals, or small groups, over the years. In the late 1940s, Alex Green, then at Florida State University, developed a Dirac model of nucleon-nucleus interactions. He soon switched his focus, however, when his major paper on the subject was rejected by the *Physical Review* on the grounds that (to quote the letter of rejection) it was "almost entirely speculative, and that the chance that it will have something to do with physics or that it will have a beneficial influence on the progress of the subject is remote" (Green 1985).

Later, at the University of Florida, Green and several students, particularly Dudley Miller, continued to develop Dirac models for nuclear physics (Miller and Green 1972; Miller 1972). Fifteen years later, Miller's work is still cited, but, after postdoctoral positions at both the University of Maryland and the Massachusetts Institute of Technology, Miller failed to get tenure at the University of Virginia and essentially left physics. Green's funding ended in 1970 when the U.S. Department of Defense was prohibited from supporting basic research, and he turned to more applied work (Green 1985). In this period, then, the environment in nuclear physics was not supportive of Dirac models—or of people who advanced them.

Another person who advanced the cause of Dirac models in nuclear physics was Dirk Walecka, longtime professor of physics at Stanford University. In 1974 he published what is now known as the Walecka model for nuclear matter (Walecka 1974). His original interest, however, was in the properties of nuclear matter in *stars*. And so his work was as much astro-physics as nuclear physics. Nevertheless, he and several of his students are now prominent among the people working on applications of Dirac models in nuclear physics (Serot and Walecka 1986).

By all accounts the work that did the most to spark interest in Dirac models for nuclear physics was that of Bunny Clark and her collaborators (Clark, Hama, and Mercer 1983). Unlike either Green or Walecka, the Clark group was not concerned to develop a nuclear model in terms of the inter-actions of more fundamental entities. They proceeded *phenomenologically*.

DIRAC PHENOMENOLOGY

To people in the humanities or social sciences "Dirac phenomenology" may sound like an obscure school of continental philosophy. In fact it is a way of constructing models, in this case models of the nuclear optical po-tential. The Dirac equation contains a set of terms that represent a number of different potentials. Some of these potentials contain free parameters. All can be assigned weights to represent their relative contribution to the total nuclear potential. For any observable quantity, such as a cross-section, one can derive from the model, with parameters and weights left unspecified, a prediction of what that parameter should be like. If one then has actual data on that quantity, one can adjust the parameters and weights to obtain a "best" fit of the data. Here one does employ some standard statistical measure of fit, such as χ^2. The object, however, is not to judge the fit between a prediction and the data, but to fit a model to the exist-ing data.

The most general Dirac equation has five different potentials. Clark's model uses only two of these—those she and her collaborators thought likely to be the most important. One of these, U^s, transforms as a scalar quantity; the other, U^v, as a vector. Other standard potentials, like the spin-orbit potential, appear naturally as consequences of these two.

The result of their curve-fitting programs was that both potentials were very large, on the order of several hundred MeV! But U^v is positive, thus repulsive, while U^s is negative, thus attractive. The difference is a mildly negative potential, about 20 MeV, very much like the potential in standard Schroedinger models. Figure 7.8 shows an equation describing this model and a graph of the two potentials. These should be compared with the Schroedinger model pictured above in figure 7.2. The fascinating thing is that although the *resulting* potentials are very similar, the *component* po-tentials are very different. The nature of the interaction between the incom-ing proton and the target nucleus is, therefore, fundamentally different in

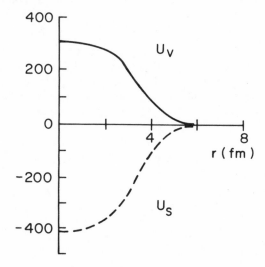

$$\{ \vec{a} \cdot \vec{p} + \beta(m + U_S) - (E - U_V - U_C)$$
$$- i\beta\vec{a} \cdot \hat{r} U_T \} \psi = 0$$

Figure 7.8. A Dirac equation with a graph of the two most important relativistic potentials.

the two models. The catch, of course, is that, before roughly 1982, almost no one believed that those large potentials could be real.

EMPIRICAL ADEQUACY IS NOT ENOUGH

Clark did not construct phenomenological models of only the Dirac optical potential. She also constructed the corresponding Schroedinger models for comparison. With only cross-section data there is little difference. To see a meaningful difference, one needs a spin-dependent observable, like polarization or spin rotation. In the early 1970s the only such data that existed were measurements of the polarization for protons scattered from helium at several different energies. For these data Dirac phenomenological models permitted a far better fit to the data than did similarly constructed Schroedinger models. These findings were presented at a number of conferences in the late 1970s, but no one paid much attention. As Clark herself later put it (in a taped interview with me), this earlier work "hadn't been looked on with a great deal of favor."

Nor were things much improved when Clark and company turned their attention to the new data coming out of LAMPF beginning in 1979. For the cross-section and polarization of protons on a calcium 40 target, at both 800 and 500 MeV, the Dirac phenomenology gave a substantially better fit than any corresponding Schroedinger models. But this still did not convince many people that there was anything to the Dirac approach.

If, as van Fraassen (1980) maintained, empirical adequacy is the name of the game, this work should have been well received. No one seems to have questioned whether the Dirac phenomenology produced superior fits to the existing data. Nor can one take refuge in the standard empiricist reply that the Schroedinger models must have been *simpler*. In fact, it is the Dirac model that is usually touted as being simpler, or "more elegant." And phenomenological models of both types typically employ roughly the same number of adjustable parameters.

Among the reasons given for not taking Dirac phenomenology very seriously is the typically *realist* objection that existing theories of the microscopic constituents of the nucleus do not allow for the existence of the required large positive and negative potentials. These potentials may make it possible to reproduce the data, but there is no reason to believe that they are really there.

A second reason is simply that the phenomenological approach uses roughly a dozen adjustable parameters. The belief is widespread that with that many free parameters, one could get a pretty good fit with just about any model. Ignoring the fact that Dirac phenomenology provides a better fit than Schroedinger phenomenology, this objection reveals a typical satisficing rationale. Figure 7.9 repeats the decision matrix for this problem. Here the state "S fits best" means that the Schroedinger model better represents the real nuclear potential than does the Dirac model. If a good fit to the data is fairly likely no matter which model better represents the real nuclear potential, there is no satisfactory decision rule based on whether either model fits the data. One has no assurance of being likely to end up

	SCHROEDINGER MODEL FITS BEST	DIRAC MODEL FITS BEST
CHOOSE SCHROEDINGER MODEL	CORRECT	INCORRECT
CHOOSE DIRAC MODEL	INCORRECT	CORRECT

Figure 7.9. The decision matrix for a choice between Schroedinger and Dirac models.

with a correct outcome. And so the fact that either type of model fits the data provides no basis for a decision one way or the other.

THE PREDICTIVE SUCCESS OF DIRAC PHENOMENOLOGY

Once the parameters of a phenomenological optical model are fixed by fitting the cross-section and polarization, one can then use the resulting model to *predict* what the spin rotation parameter should look like. This Clark did for both Dirac and Schroedinger models. The problem, of course, is that before 1981 no reliable spin rotation data existed. The 500 MeV data on calcium 40 were slow in coming out, but when Clark did get a hold of them, some six months before publication, the fit was very good. She and her collaborators immediately began working in earnest.

They achieved the expected result that the Dirac equation would yield a better fit to all three experimental quantities (cross-section, polarization, and spin rotation) than the Schroedinger equation. But what was truly exciting to these researchers was that when they determined the free parameters in the model using only the cross-section and the polarization, the resulting Dirac model yielded a spectacularly good prediction for the spin rotation. The corresponding Schroedinger phenomenological model yielded poorly fitting predictions for spin rotation. The Dirac curves, first published in 1983, but available to colleagues in preprints beginning in late 1981, are shown in figure 7.10.

The results were similar if one set the model parameters by fitting the cross-section and spin rotation and then "predicted" the polarization. The Dirac phenomenology yielded very good fits for the predicted observable, whereas predictions based on a corresponding Schroedinger model were poor. Here Clark and company had a bit of received wisdom on their side. Everyone seemed to think that the two spin observables (P and Q) ought to be connected and that a good model should reveal that connection.

Although a few people around the country saw Clark's results late in 1981, her calculations were first presented publicly at a conference on "The Interaction of Medium Energy Nucleons in Nuclei" held at the Indiana University Cyclotron Facility in October 1982. Clark knew they had finally hit it. She was, she later recounted, "very excited" and "very pleased" with the reaction.

THE RESPONSE

All the physicists I have spoken to at length about the topic agree without any doubts that one of the most important sources of increased interest in Dirac models was the *predictive success* of Clark's Dirac phenomenology.

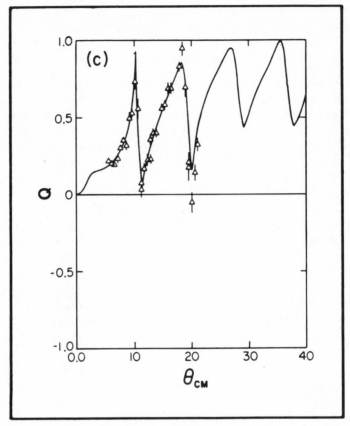

Figure 7.10. A comparison of Clark's predicted curve for Q with the existing data. Reproduced from Clark, Hama, and Mercer (1983, 271).

One recollection, by a young theoretician in early 1986, is worth quoting. I asked him why interest in Dirac models increased so dramatically beginning in 1983.

> Well, I think the most, as least as far as the scattering experiments go, the most important thing was Bunny Clark's work, because she started in the old days. I mean, she'd been working on it [in] '76, '77, I don't know exactly, but way back, and uh she was working on it, how can I describe? She wouldn't be deterred from doing this, OK, and she wasn't really doing it in a way that was the most convincing that it could be, but she was doing something

that was well defined and that was useful. And at the IUCF work-shop . . . in '82, this was after this data came out, . . . and this, there was all the same discussion by nonrelativistic-type calcula-tions of "This doesn't work," and "We have to do this," and "We have to stand on our heads, and, and try to figure out all funny little ways to, to be able get the spin rotation data to work." And then Bunny showed her calculations where she took the other two existing pieces of data, made them work, and the spin rotation function came out perfect. So I mean, it was, people were really shocked. . . . I mean it was just like, wow!

Most of my other respondents were more restrained, but the message was similar. And the same message comes through in more sober, pub-lished commentary. In a paper given at the same workshop in which Clark made her presentation, one of the Texas group who had been in corre-spondence with Clark concluded his otherwise nonrelativistic paper with a short section on "relativistic approaches" (Ray 1983, 140, emphasis in the original):

The significant feature of these calculations is that given a good fit to the cross section and *Ay* (or spin rotation) the spin rotation (or *Ay*) is *correctly predicted by the Dirac phenomenology.* Such is certainly not the case in the above discussed non-relativistic KMT approach nor in standard phenomenological optical model analy-ses based on the Schroedinger equation.

Note especially the expressed symmetry between predicting spin rotation, given the cross-section and the polarization (analyzing power), and pre-dicting polarization, given the cross-section and spin rotation.

Again, in a later review of the experimental literature, another member of the Texas group, referring to the work of Clark and associates, wrote (Hoffman 1985, 46, emphasis in the original):

A very interesting result of this work was the fact that the spin rotation function was *correctly predicted* whenever the cross sec-tion and analyzing power data were fit by minimizing χ^2. This situation was quite different from that encountered using the stan-dard Schroedinger equation approach. . . .

These experimentalists obviously think the ability to predict data is im-portant.

Further, in a report published in the March 1984 issue of *Physics Today,* titled "Relativistic Treatment of Low-Energy Nuclear Phenomena" Clark's work was described as follows:

> Clark's analysis of the 1981 LAMPF data made converts to the Dirac formalism. At a workshop on the interaction of medium energy nucleons in nuclei, held at Indiana University in the Fall of 1982, she reported that she and her colleagues, Shinichi Hama (Ohio State) and Robert Mercer (IBM), had been able to achieve excellent phenomenological fits to the LAMPF 500-MeV cross-section and analyzing power data where the non-relativistic approach has failed. But what really made people sit up and take notice was the ability of these fits to *predict* with startling accuracy the spin-rotation functions, which had become available only after a new focal-plane polarimeter, capable of yielding double scattering measurements, had been installed at LAMPF. Once again, the corresponding predictions from non-relativistic fits were badly off the mark.

Note that this was written for the general physics community, not for specialists in nuclear physics.

Finally, standard sociological measures clearly point to the predictive success of Dirac phenomenology as an inspiration for further work on Dirac models. Since the beginning of 1983 dozens of articles on Dirac approaches to nuclear physics have appeared. Virtually every one of those papers cites the work by Clark and associates. The citations generally appear in the opening paragraphs of the paper where, by convention, one briefly sets out the motivation and background for the rest of the paper. This of course does not mean that the authors were *personally* motivated by Clark's results. But it does mean that they thought this an appropriate article to cite in support of the importance of working on Dirac models. Moreover, this form of citation reflects a judgment about what the audience will judge to be appropriate motivation. In sociological terms Clark's studies have certainly had much to do with legitimating further work on Dirac models.

Why Successful Predictions Matter

So far I have demonstrated *that* the predictive success of Dirac phenomenology was a major factor in the surge of interest in Dirac models for nuclear physics since early 1983. But I have not yet attempted to explain *why* this should be so.

The doctrine that successful predictions are particularly important in the choice of a scientific model can be traced back to Greek astronomers in the late Hellenistic period. It figured in discussions of scientific method in the seventeenth century, was debated by Mill and Whewell in the mid-nineteenth century, and was vigorously championed by C. S. Peirce at the end of the century. It turned up again recently as the primary criterion for progress within Lakatos's methodology of scientific research programs. This is no place to review so long a history. Here I will concentrate on the role of successful predictions in decisions to pursue relativistic models of the nuclear optical potential. The more general applicability of the account should be obvious.[2]

Some philosophers and Mertonians might claim that I have merely uncovered the operation among nuclear physicists of a general methodological rule, or norm. The rule is, roughly, that a model that can correctly predict part of the data is preferable to one that is constructed by empirically fitting all the data. The claim would be that the existence of some such rule adequately explains the words and actions of these physicists. What remains to be done is to show how the existence of such a norm functions to further the general aims of the scientific enterprise.

Now, I would not deny that such a norm exists, or even that it is sometimes cited by scientists themselves as a rule to be followed in particular cases. I just do not think that the existence of such a rule goes very far toward *explaining* the beliefs or actions of scientists. Nor do I think it promising to seek the function of such a norm relative to some supposed general aims of science. Rather, I think we should see the norm as merely a by-product of the underlying decision strategies of scientists. In particular, I think this norm can be explained as a natural consequence of following a satisficing strategy for making decisions about the match between models and the world.

WHAT COUNTS AS A PREDICTION?

In everyday parlance, to predict something is to say, in advance, what will happen. In referring to past predictions it is often implied that the prediction was successful, that is, that what was predicted later turned out to be correct. But this does not seem to be the meaning of 'prediction' in the present context.

As a matter of fact, Clark and her colleagues had produced plots of the spin rotation function *before* the relevant spin rotation data became available. But this fact seems to have played no role in anyone's thinking about the case. The way the participants talk and write about it, "predicting" the spin rotation merely means deducing it from a phenomenological model

—one that was constructed using data on *other* observable quantities, namely, cross-section and polarization. This is particularly clear from the quotation above in which the author speaks also of "predicting" the polarization from a model constructed using the data on cross-section and spin rotation. This kind of predicting would be impossible if "prediction" depended on the temporal order in which the analyzing power and spin rotation data became available.

In what follows I shall use the term 'prediction' in roughly the sense used by these physicists. I do so not because I think there is no role in scientific decision making for predictions that are successful in the full temporal sense of being correct before the event. It is just that temporal order played no role in this case.

WHAT WERE THE OPTIONS?

I shall argue that the importance attached to the predictive success of the Dirac phenomenology can be explained if we assume that scientists' decisions about Dirac models fit a satisficing strategy. To apply a satisficing model, we must first define just what the options were.

The concluding paragraph of the written version of Clark's presentation at the Indiana workshop consisted of the following single sentence (Clark, Hama, and Mercer 1983, 278):

> In conclusion, it appears that Dirac phenomenology can play a useful role in the relativistic description of nuclei.

In Laudan's terms she was saying that Dirac phenomenology is worth *pursuing*—not a surprising conclusion from a group that had been working on it for a number of years. By implication she was saying that "the relativistic description of nuclei" was worth pursuing.

This seems to me a weaker conclusion than what was in the minds of most people at the time. The later (March 1984) *Physics Today* report put the issue more realistically:

> How does one decide whether the potentials deduced from these phenomenological fits with a fistful of adjustable parameters have any serious relation to reality?

And a still later review by a theoretician (Picklesimer 1985) used phrases like: "If there is any 'truth' embedded in the previous RD [relativistic Dirac] successes. . . ." Picklesimer explicitly formulated the questions: "Is the RD approach really so much better than its NR [nonrelativistic Dirac] counterpart that one should believe it incorporates an extra, essential piece of physics?" and "What is the basic physical mechanism which the Dirac approach adds to the NR approach?"

My interpretation is that the basic question on peoples' minds in 1982 was, roughly, whether these polarized beam experiments really were detecting relativistic potentials. The data were too sparse to decide with certainty that they were or that they were not. But one could decide whether they *might* or *might not* be detecting genuine relativistic mechanisms. Here 'might' must mean more than that it is just possible. It must mean that it is plausible, or even likely.

And so the alternatives in everyone's implicit decision matrix were, I suggest, something like: "Conclude that these experiments might be detecting relativistic potentials" or "Conclude that these experiments probably are not detecting relativistic potentials." The corresponding states of the world would then be: "These experiments are detecting relativistic potentials" and "These experiments are not detecting relativistic potentials." With this understanding of the options and states, the implicit decision matrix is basically that already exhibited in figure 7.9.

One virtue of this way of understanding the options is that the *general* pursuit questions are answered automatically. If one concludes that relativistic potentials might be involved, it follows, given the context at the time, that it is, in general, worth pursuing relativistic models and related experiments. If not, then it is not. This still leaves open the question for any individual whether he or she should actively pursue relativistic studies.

VALUES

I have argued that there is always a generalized epistemic value in being right, for example, in concluding that there might be relativistic interactions if there indeed are, and not if there are not. But most individuals will have other commitments that lead them to prefer being right one way rather than the other. The same data may therefore lead to different decisions by different people. For the moment I will concentrate on the general structure of the decision problem as determined by the epistemic values alone. Later I will consider the impact of other values. This separation is for analytical, and expository, purposes only. My account does not assume that scientists themselves make any such distinction in their own decision making.

THE ROLE OF PREDICTIONS IN THE DECISION

To appreciate the role of the successful prediction in this case one must realize that polarization and spin rotation are *independent* quantities. Both depend on the nature of the interaction between the polarized proton and the nucleus; but the dependence is different in the two cases, and so there is no necessary correlation between them. Their relationship is a contingent feature of any nuclear model.

From this theoretical fact, known to all, it follows immediately that finding spin rotation data agreeing with a Dirac model constructed using only polarization data would be very improbable—if the real interaction were correctly described by a corresponding Schroedinger model. And vice versa. According to the satisficing model of scientific decision making developed in the preceding chapter, this low probability provides the basis for using the intuitively obvious decision rule: Go with the Dirac model if it correctly predicts the spin rotation data; otherwise, stick with the Schroedinger model. Given the known model-based probabilities just noted, following this simple decision rule makes it probable that one will end up with one of the epistemically preferred outcomes.

The overall situation is very different if the competing phenomenological models are constructed by fitting *both* sets of spin data. One still finds that the standard Dirac fit is better than the standard Schroedinger fit. But there is not much basis for judging that the good fit of the Dirac model to both sets of data would be improbable if the real interaction were basically nonrelativistic. The superior overall fit of the Dirac phenomenological model could well be mainly due to having different types of potentials to play with, whether or not these potentials represent real interactions.

To put it slightly more technically, suppose that the interactions are really nonrelativistic. If the parameters of the Dirac model are determined using both sets of data, there is still some chance that the second set will not fit the model very well. But if one determines the parameters of the Dirac model using only the first set of data, the chances of then getting a poor fit to the second, independent data set must increase. The same is true of the Schroedinger model if we suppose that the interactions are relativistic. And so the test is more discriminating if the models are constructed using only the first set of data than if both sets are used in setting model parameters. Consequently, the success of the Dirac approach in predicting the spin rotation data provides a stronger basis for deciding in its favor than does its better overall fit if both sets are used to determine model parameters.

A more informal way of stating this difference suppresses reference to model-based probability judgments. If the model can produce information on *two* measured quantities when supplied information on only *one,* we can conclude that the model itself *contains* some real information about the process generating the measured quantities. Being able to reproduce information on two quantities when supplied information on both to start with shows only that the model can *accommodate* both sorts of information.

But this simple formulation only suppresses the probability judgments, it does not eliminate them. For even here the question arises whether the model is supplying real information or merely by chance yielding the additional measured values. The model-based probability judgments are required to reduce this chance.[3]

Further Evidence

I have just argued that the general reaction of nuclear physicists to the predictive success of Dirac phenomenology can be explained if we suppose these physicists are following a satisficing strategy. But one might persist in asking whether these scientists are really responding to the implicit decision problem as I have explained it or simply following a rule they have learned.

One cannot simply ask a scientist "Are you a satisficer?" because few would understand the question. Nor does it seem very promising to explain satisficing strategies to a scientist before asking this question. The overwhelmingly likely answer would be an ambiguous "Yes, it might be something like that." And standard research on human judgment does not proceed in this manner. The best I can offer is a few excerpts from interviews in which I asked scientists to explain, in their own words, why they believed what they did, why they were impressed, or not, with the predictive success of Dirac phenomenology, and why they did or did not change their minds. I would not pretend this evidence is definitive. At best it is suggestive.

THE RELATIVISTIC IMPULSE APPROXIMATION

Once a nuclear theorist started to take seriously the Dirac approach to nuclear physics, an obvious next step was to develop a *relativistic* version of the impulse approximation. That is, calculate the optical potential of a nucleus using the Dirac equation to represent the interaction between the incoming proton and the assumed "quasi-free" nucleons in the nucleus. One group that did just that consisted of Steven Wallace, a well-known nuclear theorist at the University of Maryland; Jim McNeil, his former student; and, later, Jim Shepard, a third collaborator, from the University of Colorado. I want to show, first, how the predictive success of Dirac phenomenology influenced the thinking of the Wallace group, and, second, that the same sort of satisficing strategy was present in their judgments that their own relativistic impulse approximation was correct.[4]

"That Is Breathtaking Agreement"

All three started out believing there was nothing to Dirac models in the context of medium energy nuclear physics. As one put it: "I was a disbeliever, a serious disbeliever." Wallace, in particular, had been working on proton helium scattering in the mid-1970s at the same time as Clark. He recalled seeing Clark give presentations at meetings of the American Physical Society, and his response was "Why is she doing this?"

Then Wallace and McNeil saw the LAMPF spin rotation data for calcium at 500 MeV, and "that's when the bubble burst." One of them met Clark at a meeting in the spring of 1982, some six months before the Indiana workshop.

> We sat down and she showed me her pictures of Q, and they [the curves] went through every single data point. And I said, "Can you fit that?" and she said, "No, we fit P." She fit only polarization and she made an absolute prediction of Q, ..."

Not only was it important that Clark predicted Q, but the *quality* of her predicted fit to the data was also very impressive:

> I mean, its not like going through one point. I mean, this is real detailed data with lots of nontrivial structure.

This seems to indicate a belief that successfully predicting Q would have been highly improbable if the Dirac model did not capture something important. I asked the group whether they thought the good fit to Q could have been an accident.

> Of course it could have been an accident, but her agreement was so remarkable, the first thing that entered my mind I thought is that it had to have been a fit, not a prediction. That is breathtaking agreement.
>   When I first looked at it, I assumed this was simply another fit, which Bunny had been doing for years. . . . This was the first prediction she made, and it was dramatic.

The elements of a satisficing strategy seem all to be in place. The most important thing, of course, is the judgment that the success in predicting the spin rotation measurements would have been highly unlikely if there were nothing to the Dirac model. That this judgment is fulfilled when Q is *predicted*, but not when the model is fit to all the data, is also explicit.

And what did Wallace and McNeil conclude in the spring of 1982? "It smelled very good." "There's some truth here; we didn't understand it."

"We Were out to Prove It Wrong"

Neither Wallace nor McNeil were yet convinced that Clark's models were right. Wallace recalled,

> At that point it seemed to me that the big remaining objection was that you had no idea where these crazy potentials come from; there was no way to justify them.

McNeil agreed, saying, "I mean, it just can't be; it's based on phenomenology and has no right to work." The two thought that if they went back to the beginning and developed a relativistic version of the impulse approximation they could show how it could work without the large positive and negative potentials Clark was getting. "We thought we were going to kill all this Dirac stuff. . . . We were out to prove it wrong."

But they did not make much progress though the summer, and what they did find was ambiguous. Then came the October workshop at Indiana. Clark's data were no longer news to Wallace and McNeil. But they were impressed with some theoretical calculations presented by Jim Shepard. He had done some relativistic calculations of the nucleon-nucleon interaction that invoked scalar and vector potentials and included the creation of mesons. His findings convinced Wallace and McNeil that those large positive and negative potentials might indeed be necessary to account for the spin rotation data.

> We immediately decided that Jim Shephard was a collaborator. I mean, it was, it's a very simple argument. As soon as you see it, bingo, that's got to be it. . . . So the argument was so convincing that we merely sort of shut up and said, OK, let's go home and do this. . . . So we went back and this became front burner stuff.

McNeil soon discovered the source of their earlier failures in their computer codes ("That's what happens when you divide by zero"), and by some time in November Wallace had developed the relativistic impulse approximation (RIA) to the point that it provided a convincing theoretical basis for those "crazy potentials." Wallace and McNeil had become, in their own words, "converts."

"We're Gonna Be Rich and Famous"

Meanwhile, Jim Shepard had gone back to Colorado and began working furiously on calculating polarization and spin rotation functions using an RIA. Wallace's work, which as an official coauthor he saw immediately, helped him a great deal.

> Anyway, we were working ahead on this, and we knew exactly what to do. And finally, I guess it was like a week before Christmas, the first calculations came out. There were a number of simplifications, you know, I just wanted to see if it was going to be on the same sheet of paper. And so we're sitting there, and there's a plotting machine. None of us know what the numbers are going to look like. And it plots the data, and it plots the curve, and they're real, real close. I knew that was it. I said, "Dave, we're gonna be rich and famous." And the amazing thing was that there were a number of simple approximations that could have been relaxed. Each time we relaxed the damn approximations, it got closer and closer. This happened over a period of a week.[5]

Shepard had to make an official trip to Washington shortly before Christmas and stopped off to see Wallace. They were very pleased with the progress.

> Now our relativistic [calculation] was going right through all the data points. And the other thing is that the potentials, numerically speaking, were totally consistent with what Bunny had found phenomenologically. So that, in my opinion, I think the opinion of lots of other people justified her phenomenology, put it on a sound basis.

Shepard's final curves are reproduced in figure 7.11. They have been so often reproduced in the literature that they have become something of a symbol for the Dirac approach.

A Better Decision

The decision to accept the relativistic impulse approximation is even better than that to accept Clark's phenomenological Dirac model. Constructing an RIA does not require fitting *any* of the spin data, not polarization, and not spin rotation. Instead, it incorporates completely different data about nucleon scattering and nuclear interactions. The fact that it yields correct "predictions" for *both* of the spin observables is thus even more remarkable than that a phenomenological model can predict one or the other. If one is working with a fundamentally mistaken model, it is more unlikely that one should get two observables right than that one should get one

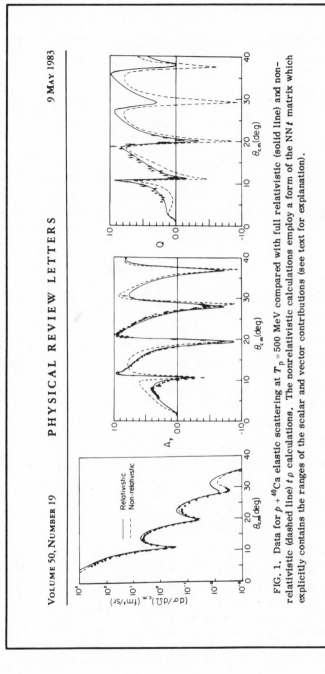

FIG. 1. Data for $p + {}^{40}\text{Ca}$ elastic scattering at $T_p = 500$ MeV compared with full relativistic (solid line) and non-relativistic (dashed curve) $t\rho$ calculations. The nonrelativistic calculations employ a form of the NNt matrix which explicitly contains the ranges of the scalar and vector contributions (see text for explanation).

Figure 7.11. A comparison of all three types of data with both a standard impulse approximation (dashed curve) and with the relativistic impulse approximation (solid curve). Reproduced from Shepard, McNeil, and Wallace (1983, 1445).

right. And so the satisfaction level for a satisfactory decision regarding the RIA could be even higher than that for a decision regarding Dirac phenomenology.

This analysis of the situation is well reflected in the following comments by another theorist intimately involved in this strand of research. I asked him if he thought these scattering experiments truly were detecting relativistic effects, and if so, why. He referred to Clark's work and also to the relativistic impulse approximation.

> The [relativistic] impulse approximation, I mean, it, it's parameter-free in the sense that all of the parameters are determined from independent, independent data, and then it just, the scattering just comes out by itself. To me that has even more impact than fitting half the data or a third of the data. . . . I mean, in other words, it's, it's completely possible that if they would have sat down to do the calculation the data would have, would have been on a different page, a different piece of paper than the curve. I mean it could have been off by three orders of magnitude. So the fact that it isn't, you know. . . .

What he was saying, in literally graphic terms, is that if the *standard* IA were right, the chances of fitting those data curves using the RIA would be tiny. Yet the RIA says that is just where the data should be. The pattern for a test of the RIA, as set out in the previous chapter, is all there.

The Design and Execution of an Experimental Test

I shall now describe a particular experiment designed and carried out by a group at the Indiana University Cyclotron Facility. At several points in the process one can see quite persuasive evidence of a satisficing strategy being employed by real scientists.

DESIGNING A SPIN ROTATION EXPERIMENT

In June of 1983, eight months after Clark's presentation at the Indiana University workshop, an IU group submitted a proposal to the Program Advisory Committee entitled "Measurement of the Spin Rotation Parameter Q in 200 MeV Proton Elastic Scattering from ^{12}C and ^{40}Ca." A section titled "Model Calculations of Spin Rotation" included a discussion of Schroedinger and Dirac phenomenological models, with explicit mention of Clark's work. The proposal presented several members each of the Schroedinger and Dirac families of models and exhibited predictions for Q, as shown in figure 7.12.

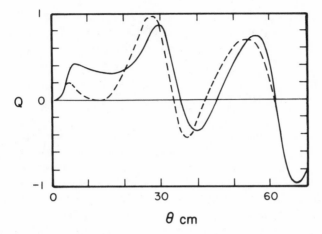

Figure 7.12. Plots of the spin rotation parameter for one Schroedinger model (dashed curve) and one Dirac model (solid curve). Reproduced by courtesy of the Indiana University Cyclotron Facility.

As one would expect, the proposal also contained considerable discussion of the experimental setup and of the particular equipment to be used, some of which was designed expressly for spin rotation experiments. The connection between the models and "experimental realities" was clearly delineated in the following paragraph (emphasis added):

> We expect the statistical and systematic contributions to our determination of Q to be less than 0.03, a number substantially smaller than the range of variation among the predictions of these various phenomenological models. Thus these models will be substantially more constrained by the new measurements, and *discriminating choices among them may become possible.*

In other words, the expected experimental error in measurements of Q is plus or minus 3 percent, but the differences between the values of Q predicted by the various models are "substantially" greater. And so "discriminating choices" may be possible.

It requires almost no interpretation to fit the description of this experiment into the satisficing model of an experimental test developed in the previous chapter. If one of the Dirac family of models is correct, the data points are unlikely to vary by more than 3 percent from the corresponding prediction for Q as a function of the scattering angle. Similarly, the data points are highly unlikely to vary enough to fall within 3 percent of the curve for Q predicted by models in the Schroedinger family. And vice versa. If one adopts the obvious decision rule (Choose Dirac if the data fit the Dirac prediction; choose Schroedinger if the data fit the Schroedinger prediction), one has a basis for a "discriminating choice." That is, one has a satisfactory decision problem.

DECISIONS ON THE RUN

I observed parts of one run of this experiment. On about the fourth day of a six-day run, I found some members of the research group, as usual, around a table by the computer in the data acquisition area. The data from the detectors go directly into a computer and onto a tape. But a small subroutine gives a quick and dirty calculation for Q so that the group can maintain some immediate control over what they are doing. The experimental values of Q obtained so far had been plotted on standard graph paper. Scattered over the table were a half dozen or so graphs of predicted values of Q for various models, some relativistic, some not. The actual values were compared with the various predictions by superimposing one graph over the other. Indeed, a light table in one corner of the room facilitates such inspections. Figure 7.13 shows two of the predicted curves, one relativistic and one not, together with the data points.

My own model of an experimental test predicts that this research group would be most interested in measuring values of Q at scattering angles between 5 and 20 degrees. That is where the differences between the relativistic and nonrelativistic predictions are generally the greatest. But that is not what we see here. There is one measurement at 10 degrees, but a half dozen at angles greater than 25 degrees. What is wrong?

In studying the actions of scientists, there is no substitute for being able to ask informed questions. Why had they at this point not taken more measurements in the 5 to 20 degree range? In fact, they had several good reasons.

Most important, measurements inside 20 degrees are much more difficult to take than those outside 20 degrees because at the smaller angles, the detectors are very close to the path of the primary beam. The experiment is designed to detect protons elastically scattered at a given angle. But the beam hitting the target produces all sorts of other reactions, including inelastic reactions resulting in the ejection of protons and other forms of radiation. This extraneous radiation is most pronounced in the forward direction, that is, along the path of the primary beam. At small angles, therefore, the detectors are being bombarded with all this extraneous radiation. By means of an elaborate sequence of coincidence and anti-

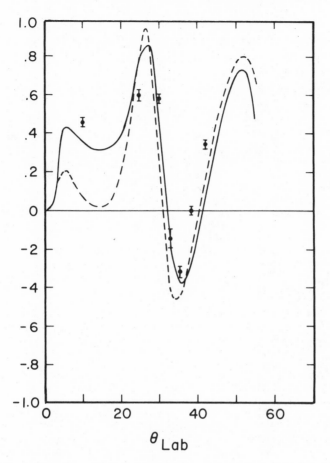

Figure 7.13. A comparison of the initial 200 MeV spin rotation data with one Schroedinger model (dashed curve) and one Dirac model (solid curve). Reproduced by courtesy of the Indiana University Cyclotron Facility.

coincidence, the detecting system attempts to sort out the desired elastically scattered protons from everything else. But if there is too much other stuff hitting the detectors, these electronic sorting devices are overwhelmed and misrecord the desired events.

Although there are various ways of coping with this problem, in the end the only way significantly to reduce overloading is to lower the intensity of the incoming beam. This means there will simply be fewer protons interacting with the target and less "noise." The trouble with this solution, of course, is that there are then fewer of the desired reactions as well. Thus, the time required to obtain enough relevant events to keep the statistical margin of error within the desired bounds goes up. And the time per data point increases.

A similar problem arises at very large scattering angles, say, beyond 50 degrees. For any incoming proton the probability of being scattered is less at a larger angle than it is at a smaller angle—just as for billiard balls. The counting rate therefore decreases for measurements at larger angles, again increasing the run time per data point. As far as speed and ease of taking data is concerned, the preferred angles are in the range between 25 and 50 degrees.

A second factor influenced the decision to take the first data at these experimentally easiest scattering angles. This run was the first with all the newly constructed equipment in place. The team decided to do the relatively faster and easier data points first in order to build their "confidence" in the new equipment. The more difficult points were saved until after the team could be confident that the equipment was working as intended.

Finally, measurements at the smallest possible angles, down to roughly 6 degrees, require a change in the physical configuration of the reaction chamber and detectors. This means breaking the vacuum in the reaction chamber, making the required changes, and then reestablishing vacuum before data taking can resume. The process takes perhaps four hours, half a shift. As a result the team would want to save these measurements until last to preclude wasting time getting things back into the more normal configuration. But even then the team has to judge carefully how long it will take to acquire the data. And they will have to gamble that nothing else will happen to interfere with data taking—like a breakdown in the cyclotron itself.

In short, once one understands the more detailed circumstances of the experiment, one sees that this group of experimentalists was after all doing what my satisficing model of experimental tests would predict. They did regard the points in the 5 to 25 degree region as the most important, and they organized their data taking to obtain the most reliable possible values of Q at just these angles.

Note also that even while working on the easier angles, the group seemed to be following a satisficing model. They did not just measure values of Q at regular intervals, say, every 5 degrees. Rather, the first six data points were taken at scattering angles of 49, 42, 38, 36, 33, and 31 degrees, respectively. What determined this irregular pattern? Obviously, they took account of the fact that one of the distinctive differences in the predicted curves for Q is the shape and location of the second minimum at about 35 degrees. These researchers were trying experimentally to distinguish the models in just the way a satisficing strategy would recommend.

A final note illustrates the contingencies in scientific research. After the experimenters completed the run, they discovered that the computer program used to calculate values of Q contained an error. The "correct" values are much closer to the Schroedinger model predictions shown in figure 7.13. But this mistake does not affect my analysis of the decision making involved while this particular experimental run was in progress.

In all my talks with the physicists involved in this research, I have found almost nothing that could remotely be interpreted as evidence of Bayesian thinking. These scientists do not use words like 'probable' or 'likely' when talking about models or research approaches. One never hears a comparison of models in terms of "odds." It is not that I have edited out such comments. They simply are not there. It does not follow, of course, that my own satisficing account is necessarily correct. But I think I have shown that it provides a far better explanation of crucial aspects of actual scientific decision making than does the Bayesian account.

Cognitive Resources and Scientific Interests

Most of the above discussion focuses on the *epistemic* value of making correct decisions and abstracts from other interests. That other interests might be important in this case, however, is suggested by the simple fact that not all nuclear physicists are convinced there is anything to the Dirac approach as applied to medium-energy nuclear reactions. Indeed, even proponents of the Dirac approach estimate the percentage of sympathizers in the field as a whole to be at most 50 percent. One obvious source of this difference in judgments is a difference in the values attached to the two correct outcomes in the standard decision matrix. Even if all agreed on the relevant model-based probabilities, which is hardly the case, a difference in values attached to correct outcomes could render the same choice satisfactory for some and unsatisfactory for others.

The range of possibly relevant nonepistemic values is very large, far too large to treat systematically here. But it is important to illustrate how important a role such values play in the scientific enterprise. I will therefore devote this section to examining one value whose importance emerged during the course of my research—the value of an individual scientist's *cognitive resources*. Cognitive resources naturally generate what any sociologist of science would recognize as potential scientific interests.

What do I mean by a cognitive resource? For theoreticians the models with which one is familiar are cognitive resources. But I do not mean just an abstract, intellectual familiarity. I mean the ability, even the *skill,* to work with the models, to apply them in new cases, to use them in calculations, and so on. Furthermore, as the construction of models and derivations from them becomes ever more dependent on computers, the ability to use computer codes, even to write them oneself, becomes an increasingly important cognitive resource.

Experimentalists have many of the same kinds of cognitive resources as theoreticians, though theirs tend to be deployed less extensively. In addition they have a wide range of skills necessary in working with the equipment and instrumentation involved in modern experimental science. Designing an experiment is a highly specific skill, so specific, in fact, that nuclear physicists develop reputations as experts in inelastic rather than elastic scattering, for example, or in the use of specialized kinds of detectors (Galison 1985).[6]

I shall not proceed systematically. Rather, I shall merely illustrate the role variation in cognitive resources played in the decisions of several individuals in my study. The data again are from my own interviews.

Clark

Bunny Clark provides the most interesting case because she started using Dirac models to describe proton scattering at a time when almost no one else in the field thought they were at all relevant. And she approached the problem phenomenologically when, as is clear from previous quotations, many theorists placed little value in a phenomenological approach. Why was she doing these things? The keys to the answer were in her very first response to my questions. I noted that everyone had told me she had been in the Dirac business "way back when." "So," I said, "tell me a little bit about 'way back when'."

> Well, we came at the problem from electron scattering, actually. In electron scattering you always use the Dirac equation, and there aren't any of the conceptual problems that people have with using the Dirac equation for proton scattering because for electrons you know what the interaction is . . . and for years I worked with the people at Stanford. . . . We were solving the Dirac equation for electrons—and I had gotten married before I had finished graduate school, . . . and went with my husband who had finished his Ph.D. up to the state of Washington. I got a job with General Electric in a very small group that was doing, essentially, calculations—was supposed to be theoretical physics—and that's where I learned how to program.

Thus, long before she completed her Ph.D., Clark had acquired two cognitive resources essential to pursuing Dirac phenomenology. She had learned how to work with the Dirac equation in the context of electron scattering. And she learned to program computers. The latter skill is essential because there is almost no way to determine a dozen parameters by fitting a curve except by a long, iterative process that requires many calculations and recalculations.

Clark finished up a master's degree in physics, and then, about 1961, moved with her husband to Detroit, where she "walked in off the street at General Motors Research Laboratories" and landed a job that permitted her to spend some time analyzing electron scattering data. She also made trips to the University of Illinois to work with people there who did electron scattering. After a while she decided to resume her graduate education at Wayne State University.

> So I started graduate work, but I'd been doing research all the time, and so it was, I guess, pretty clear I didn't have to take too many courses, I could just kind of start. And so we were trying to look at a problem which would be related in some way to the stuff that I had been doing before.

Here the connection between the choice of a new problem and existing cognitive resources is explicit. She continued:

> And there was some new data which was extremely puzzling that . . . measured protons scattered from helium. . . . And there were very great difficulties in understanding some aspects of this data from standard things that people thought they should be able to understand. Multiple scattering theories weren't working well, stuff like this, and we decided that we'll look at that. And so this was at one GeV, so that was pretty high energy, so it was pretty clear you want to use a relativistic wave equation, and I said, "Well, why don't we use the Dirac equation?"

Obviously, Clark's approach to this new problem was as much determined by her existing cognitive resources as by the problem itself.

Around 1970 the couple moved again, this time to Ohio State University, where Clark got a job as a research associate in the department of physics. Once again she had access to the computer time necessary to continue research on her dissertation. About that time some data on the polarization of protons scattered off helium became available. For the first time Clark faced the problem that her attempts to fit the data with a Dirac model yielded "these huge potentials." One person she showed them to recommend she look at Dudley Miller's (1972) paper. That gave her some comfort. So did Walecka's "lovely" (1974) paper. But she was not in correspondence with either man. There was at that time no genuine "Dirac model research tradition" in nuclear physics.

By the time the "beautiful data" from LAMPF on spin observables for calcium and lead appeared in 1978, Clark had all the cognitive resources necessary to do what she proceeded to do. There was no question but that she would pursue Dirac phenomenology. And certainly she had a strong interest in the decision theoretic outcome in which the Dirac approach is chosen as the best. Nature did not have to cooperate, of course, but when she did, Clark hardly had to be forced to decide that the Dirac approach might be right after all.[7]

Bunny Clark herself is implicitly well aware of the role of cognitive resources, as well as of other interests, in determining why people do what they do.

> Well, some people do what they think is fashionable to do because the field is very competitive, and if you're young you'd better do things that have high visibility and are guaranteed publication success rate in many cases, or otherwise you may not be able to stay in the field. That's too bad. . . . I know why I do what I do. I like what I do. I like it, and I enjoy doing it, and I'm good at it. So that's—I mean, other people are good at other things. Some of those things I couldn't do at all. That's why I do what I do.

ANOTHER THEORIST

Now consider the case of a younger theorist who completed his Ph.D. in 1979 and now has a number of publications exploring theoretical aspects of the Dirac approach. I asked him how it was he began working on relativistic models of the nucleus. The general form of his answer was predictable:

> Well, it was part of my thesis. I mean, I was, it was suggested to me as a topic by my thesis advisor. . . . In fact it's funny because, this was in the last year of my graduate work and, it was October of this year and I had already done two things and I was sort of, I went in to see [my advisor] . . . and said, "Well I'm pretty much done, aren't I?" You know. And, he said, "No." You know. "You might want to look at this," and at the time that meant sort of changing gears and doing a whole new thing, but, and I only had seven, eight months to really put in on it 'cause I had to be finished by June. So, you know, I sort of said, "There goes [my

advisor] again, he always wants more," and, in retrospect it was a good, a good thing, because I don't really do the stuff I had been working on, I was working on before. I switched, to this.

How, I persisted, did your advisor convince you that this was a good thing to do—apart from exercising pure authority? The influence of authority was immediately acknowledged. "Obviously," my respondent replied, "if he thinks it's a good thing to do, almost by definition that convinces me." But he then went on to give a more considered answer.

It was fairly easy to convince me, 'cause it, it contains things like field theory and, and, quantum mechanics and stuff that interests me, period. So once he said, "Go read my paper," that, that, he didn't really have to convince me much. I mean, that's why I'm still doing it, 'cause it's, it's the kind of theory that I always wanted to do, that used to be you could only do as a high-energy particle theorist, but now you can do it in nuclear physics too.

And so it is quite clear how this young man got into the business of working on Dirac models. But I pressed on for further details of how he came to *continue* working on relativistic models. As it turned out, he had tried hard *not* to continue. Although he liked working on Dirac models, he said, "In the old days [1979!] there had been a handful of people, you could count on one hand the number of people who were interested." He therefore set out deliberately to do something else and took a postdoctoral position clear across the country. He spent the year 1979–80 working with other people on other problems, but that, as he admitted, "didn't work out very well." The following year (1980–81) he was back temporarily as a faculty member in his advisor's department. There he found other students working on Dirac models, and largely through interaction with these other students, he drifted back into working on relativistic nuclear models himself. Two years later, however, he tried once again to break out.

Now in fact I had a year leave from [the university] in '82–'83. And the prime purpose of having this year leave was to go out and meet new people and do new things, and that year was essentially when all this relativistic scattering, Dirac business exploded. So I mean I went away with the sole purpose of doing something else, and I even did for six months, I didn't work on it, I did something else. . . . It was after I'd already worked for two years, and now's the time to, to really got to concentrate on spreading out here, and I worked for something else for six months and, and then it just, it was so explosive, I mean, things were just happening so fast and people were so excited that they just, you know, that, that I said, "This is stupid, I'm not going to fight the, fight the trend, I'm just going to go back, and its fun to do, and I'll just do it." . . . It was one of these things that I had to say, "Look, I really would like to keep doing this, but it's in my best interest to do something else," and that's why I tried to do something else.

The role of acquired cognitive resources in the early career pursuit decisions of this young theorist was obviously very important. The resource in question, of course, was his familiarity with relativistic models of the nucleus. At several different times he explicitly considered the option of pursuing investigations on Dirac models versus that of doing something else. The question whether the Dirac or the Schroedinger approach is correct seems to have played little role in these decisions. He was concerned with the short-term professional payoff—his "best interest." This concern appeared the same no matter which approach might eventually turn out to be correct.

Now, the first several times he faced the decision, when he finished his degree and then a little later, so few people were interested in relativistic models that any outcome that had him employing this particular cognitive resource appeared to have little professional value. I presume this means, as suggested by Clark's comment, there would be some difficulty publishing papers on the topic and there would be little professional recognition even if they were published.

Then, suddenly, lots of people were very interested in Dirac models of the nucleus. And there he was in possession of a relatively rare cognitive resource. Working on Dirac models promised a clearly positive short-term professional payoff whether or not the Dirac approach turned out to prevail. It seems also to be the case that he genuinely enjoyed working on relativistic models. The Dirac option was now satisfactory. And he took it.

AN EXPERIMENTALIST

My final example is the experimentalist at IUCF who directed the spin rotation experiments described earlier. He completed his Ph.D. in 1975 working on low-energy interactions using spin-polarized deuterons. After two postdoctoral positions he joined IUCF as an associate research physicist in 1979. He still describes his major research area as "nuclear reactions with polarized ions."

Already in 1975, therefore, this experimentalist was in possession of an essential cognitive resource for pursuing experimentation relevant to the

Dirac approach. He was an expert in working with polarized beams and in measuring spin variables such as polarization and spin rotation. The difference between a beam of deuterons and a beam of protons is relatively minor.

He first heard of Bunny Clark's work in the early spring of 1981 from colleagues at IUCF. And he got a firsthand look at her work when she visited the facility later that spring. That was a year and a half before the workshop in which Clark presented her "predictions" of the spin rotation parameter. Even at this early stage this experimentalist was intrigued enough to want to know more.

> The reason the time of the year is memorable is because I spent the Christmas holiday reading old textbooks about what Dirac had done. . . . I just had forgotten from my course work as a graduate student lots of mechanical things that one has to know about this equation in order to make sense of the bits and pieces and how those fit together.

And he was pleased with what he was learning.

> This was an entirely different way of looking at elastic scattering. It wasn't simply the old stuff repackaged, which is what you mostly get. It was completely orthogonal to a lot of the old stuff. And so there were enough new pieces in it that I was intrigued by it, and that's why I decided to pursue it.

What he omits to mention, of course, is that he could hardly have made this decision if he did not already possess considerable experience—a cognitive resource—with spin-polarized ion beams.

But, he noted, "There was something else that was happening at roughly the same time." And this turned out to be equally important for his later work.

> Osaka has a cyclotron, and they were trying to measure exotic spin components by polarizing the beam prior to acceleration. And I've always watched fairly carefully what they do because they're very good technologists. . . . And I realized that was something we could do here very easily without spending any money or anything like that. And with a couple talks with [a colleague at IUCF] I realized that we were sort of on the threshold of being able to do that here.

A new way of producing a polarized beam may be more a material than a cognitive resource, but for an experimentalist, material resources are crucial.

It turns out that this particular experimentalist was *not* especially impressed with the "predictive" success of Clark's Dirac phenomenology. Not that he violated my conditions for a satisficer. He simply disagreed with the required probability judgment as to how likely that success would be if those large potentials do not exist. And this skepticism was based on a general distrust of the phenomenological approach. Referring to Clark, my respondent said:

> She does a nice job of it. But there's still, I mean, 12 free parameters there, and the trouble that I always have with those kinds of free parameters is that they're only tied in a very loose way to the underlying nuclear physics.

He admitted, however, that this skepticism "makes me stand a little bit apart from the rest of the pack." Many people, he concurred, "wanted to make it a clean case that, if the agreement is this good, for the Dirac equation, then this whole approach must be right." He was unwilling to make that leap. Nor was he much impressed with the original relativistic impulse approximation developed by the Wallace group. They treated only one target, calcium 40, and one energy level, 500 MeV. "If this is right," he insisted, "it ought to work at more than one place."

Nevertheless, his caution did not prevent him from using Clark's work as part of the justification for his own experiments on calcium 40 at 200 MeV—as we saw in the previous section. This suggests that the main factor in his decision to pursue those experiments was simply that he possessed the cognitive (and material) resources to do experiments that then seemed of great interest to a broader community of nuclear physicists. Some results of that series of runs, completed in mid-1985, are shown in figure 7.14. This figure shows a comparison of the data with one Dirac model and two different Schroedinger models. These data provide no firm basis for choosing one model over the others.

The problem, as this experimentalist sees it, is that better developed Dirac models have been matched by better developed Schroedinger models.

> See, by the time you've gotten all of these things to the same level of sophistication, the differences begin to disappear. All models are, in essence, working. . . . I'm persuaded at this point that the elastic scattering experiments will not decide. . . . At this point it's really on hold. If the question is "Who's right?" nothing that

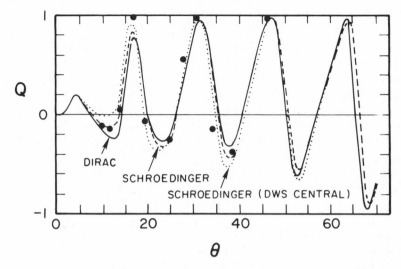

Figure 7.14. The final comparison of 200 MeV spin rotation data with two Schroedinger models and one Dirac model. Reproduced by courtesy of the Indiana University Cyclotron Facility.

I'm going to do this year or next year is going to influence that question one way or another. I see no handle that I can grab onto.

My gloss is that he no longer can see how to deploy his particular cognitive and material resources experimentally to attack the big question of which approach is right.

Does this mean he is giving up experimental measurements of spin variables for protons? Not at all.

If the question is strictly "Which equation is right?" then I think there's no reason to do any more elastic scattering. There is, however, a closely related question which often is hardly distinguished in practice, and that is "What kind of data do the people who do relativistic modeling need in order to test whether their models are working as an internal structure or not?" And the answer may be elastic scattering data. But that's really a different question.

This researcher is still (in 1987) engaged in experiments of this sort.

An Evolutionary Picture

Among evolutionary epistemologists Donald Campbell has been among the most radical in insisting that the basic mechanism in knowledge generation is "blind variation and selective retention" (Campbell 1960). Critics have pointed out the implausibility of such a thesis applied to the activities of individual scientists who generate new models or design new experiments. Their activities are surely intentional and hardly random. Yet, at a higher level of organization, there may be more to Campbell's thesis than one might at first think.

A classic empiricist view of science, as exemplified by Popper (1959a), pictures the overall activity of science as being highly intentional. Science, according to Popper, consists in the formulation of bold conjectures and deliberate attempts to refute them. An appreciation for the role of cognitive resources and other nonepistemic interests suggests a far different picture.

Models do get proposed. And experiments do get designed and carried out. That much is clear. But there seems to be little intentionality at the level of "science in general." Rather, individual theorists develop particular types of models in considerable measure because they already possess the cognitive resources to do so and perhaps because they perceive the scientific environment as being receptive to those models. Experimentalists develop experiments that test those models not primarily because they are following any general scientific norm that models are to be tested. Rather, they possess cognitive and material resources they recognize to be relevant to such tests, and they seek to employ their resources to their best professional advantage.

The analogy with evolutionary processes is striking. It looks to the uneducated eye that species are *designed* to fit their environment, or that particular species as a whole are "trying" to adapt. The truth, of course, is very much otherwise. Individual organisms are simply pursuing their own procreative interests as best they can. As a result, the species evolves to be better adapted. The appearance of higher level design or intentionality is an artifact. So also, it seems, with science.

The Future of Dirac Models in Nuclear Physics

Earlier I quoted an experimentalist who claimed that he cannot now think of any way experimentally to distinguish between the Schroedinger and Dirac approaches. It is my impression that his judgment is now shared by a majority of workers in the area. Nevertheless, at this moment no consensus exists within the nuclear physics community that the issue is dead. A relatively small group, between 50 and 75 people, are actively pushing the Dirac approach. Fewer people are actively opposing it. In the profession at large maybe as many as half have some sympathies with a few aspects of the Dirac approach, such as the relativistic impulse approximation. The majority continue on much as before.

In my view these differences in judgment about the merits of the Dirac approach do not betoken anything like "irrationality" on the part of some nuclear physicists. There is no unique, "rationally correct," decision. There are only the judgments of individual scientists. Differences in attitude regarding the Dirac approach have various empirical explanations.

MODELS AND APPROACHES

Philosophers have long debated whether "theories" are "falsifiable." The distinction between specific models and more general approaches permits an easy resolution of this long debate—at least in its descriptive form. Scientists do reject specific models on the basis of experiments. The original, nonrelativistic, Schroedinger-based model of the optical potential for 500 MeV protons on calcium 40 was one such model. Everyone now agrees that this model should be rejected, and why: It does not come anywhere near fitting the data.

But one highly specific model does not constitute a whole approach (or a whole family of models). And so rejecting this model does not mean rejecting the whole Schroedinger approach. In fact, other models in the Schroedinger family can do the job. The earliest papers describing the LAMPF data on polarization (Hoffmann et al. 1981) suggested the possible need for "nuclear medium modifications" in the standard models, that is, taking into account specific interactions among nucleons that are ignored in the standard impulse approximation. A number of such revised models were subsequently developed.

One might wish to object that these new, nonrelativistic models are ad hoc. They were introduced only after the new data demonstrated the inadequacy of the standard Schroedinger models. This is true, but it does not make them ad hoc in any sense that physicists regard as relevant. The fact is that most of the corrections that have been proposed are based on physical principles that were known all along to apply, such as the Pauli exclusion principle. The corrections were not made earlier simply because they seemed not to be necessary to account for the existing data. The new data are interpreted, within the Schroedinger framework, as showing that these "corrections" may be necessary after all.

An "Equivalence" Theorem

At a "Workshop on Current Topics in Relativistic Nuclear Physics" in June 1986, a European theoretician presented the following "equivalence" theorem: The Dirac approach with local potentials is equivalent to the Schroedinger approach with almost local potentials plus a Lorentz-Lorentz correction. 'Equivalent' here means that all the observable (in the physicists' sense) quantities one can derive from a Dirac equation can also be derived from a Schroedinger equation.

People skeptical of the Dirac approach seem inclined to regard such results as vindicating the Schroedinger approach and showing that it is useless to look for experiments to help decide the issue. Those sympathetic to the Dirac approach tend to think that this maneuver really is ad hoc. The potentials than turn up in the "equivalent" Schroedinger equation are different from most earlier Schroedinger potentials and have little basis in known physical principles. As one of my pro-Dirac respondents put it: "The repaired Schroedinger equation *is* the Dirac equation."

It will be recalled that Clark obtained far better fits with her phenomenological Dirac models than with corresponding phenomenological Schroedinger models. How could this be if there really is an equivalence? Well, it turns out that the basic form of her Schroedinger models was not the most general possible. She assumed various symmetries in the potential function that are lacking in the more general, Dirac-equivalent Schroedinger models. But at the time everyone else was making the same assumptions, and so no one questioned her on this point.

The Appeal to Simplicity

In my talks with physicists about the controversy over using Dirac models in nuclear physics, the respondents made frequent appeals to "simplicity" and "Occam's razor." But the appeals came from both sides. Those favoring the Schroedinger approach argued that the more complex, Dirac models need not be invoked if the Schroedinger models would do. Those favoring the Dirac approach argued that Dirac models are intrinsically simpler, or more elegant—everything falls out "naturally."

Philosophers who have taken seriously scientists' appeals to simplicity have generally assumed that simplicity is an intrinsic property of theories that makes them more acceptable. The task, therefore, has seemed to be to discover what it is about a theory that makes it simple, and why theories with this characteristic should be more acceptable than others (Popper 1959a). Others have taken the lack of agreement on the proper analysis of simplicity to indicate that scientists' judgments of simplicity are inherently subjective, and thus that simplicity is of no value in understanding scientific inference.

I am inclined to agree that there is no abstract property called "simplicity" that some scientific models exhibit to greater degree than others. It is just not the case that some models in nuclear physics are intrinsically simpler than others in the same abstract way that some models of evolutionary biology are simpler than others. It useless to seek a general analysis of simplicity in the abstract. On the other hand, it does not seem to me that the appeal to simplicity lacks any rationale. It does, I think, play a role in science. And that role can be explained in terms of a simple satisficing model of scientific judgment.

Nuclear physicists who claim that Dirac models are simpler than some equivalent Schroedinger models can point to definite features of those models to support their claim. It is irrelevant that these features might not generalize to other sciences, or even to other areas of physics. It does not even matter whether others agree to the description of Dirac models as "simpler." It is enough that some people will agree that the designated features are desirable, by whatever name. In the end what is important is that others will agree in *restricting* the family of Schroedinger models to those with the designated features. For once a suitable restriction is achieved, it becomes possible to apply a satisficing strategy because then not all possible experimental results will be equally probable given some model in either family. In sum, the point of the appeal to simplicity is to cut down the number of models in one or another of the rival approaches so that a satisfactory decision is at least possible.

Of course, adherents of a rival approach are not compelled to agree with the judgment that the excluded models are less simple, or that the features designated as indicative of greater simplicity are desirable. That is why appeals to simplicity are rarely decisive and have the appearance of being merely subjective.

A Possible Scenario

One possibility suggested by several of my respondents, and reinforced by many other commentators, is that the Dirac approach will gradually take over simply because it is most compatible with what is going on elsewhere in physics, particularly in *high-energy* physics. The mesons, quarks, and gluons of high-energy physics are fundamentally relativistic. If these sorts of things are in the nucleus, they must be there in low-energy interactions as well, whether they are experimentally detectable at low energies or not.

Also driving this scenario is a generational factor. Nuclear physicists who were in graduate school after 1975 have all learned the "new" relativistic theories of high-energy physics, despite their specialization in nuclear physics. For theorists, in particular, the pull toward these new theories is very strong. And since these theories are fairly new in nuclear physics, it is easier to achieve "original" results within the relativistic framework than within the older, nonrelativistic framework—even though the theory is mathematically more complex. Right now a "satisficing" young theorist will tend to pursue the Dirac approach.[8]

It might seem that such a scenario is incompatible even with a modest constructive realism. Not at all. One of the main pulls toward the Dirac approach is that it is more "fundamental"—which is the physicist's way of indicating what is physically real. It is high-energy physics that tells us what the "ultimate" constituents of the world are. Bringing nuclear physics into line with high-energy physics is to bring it closer to physical reality. That such a move also serves the short-term professional interests of young nuclear theorists does not conflict with this solidly realistic goal.

8

Explaining the Revolution in Geology

I would now like to illustrate the usefulness of the models developed in the previous chapters by applying them to the recent revolution in geology. This chapter, therefore, will be more historical in character than those preceding. But it does not pretend to be a genuine history of this episode in recent science. It is at best a sketch that illustrates how one might deploy various cognitive models in constructing a genuine history.[1]

What is now called the revolution in geology can be roughly located in the decade between the mid-1950s and the mid-1960s. One particularly interesting fact about this episode is that the models of the new geology are similar to a model developed in the years 1911–15 and debated in scientific circles throughout the 1920s. This makes possible a fruitful comparison. What was different in the 1960s that might explain why there was a revolution then and not in the 1920s? Any adequate theory of science should be able to explain this difference.

The example of geology has obviously not been chosen at random. I have been led to explore it at some depth because it seems to me a particularly clear case of a major revolution in a modern science for which the perspective of a naturalistic (evolutionary, constructive) realism is especially well suited. In particular, this case shows that it is at least sometimes possible for the data to be so strong, and the connections between the data and the rival models to be so obvious that most nonepistemic values are simply overwhelmed. Whatever their original personal, professional, or social interests, it is possible for the vast majority of a profession to be left with no satisfactory option but to accept the new models as the best available representations of the world.

Contractionist Models

Anyone still inclined to think that scientific theories are best described as (implicit?) axiomatic systems would be well advised to consult some standard texts and treatises in geology. There it becomes quickly apparent that any attempt to reconstruct geological theory as a single axiomatic system would be as difficult as it would be fruitless. Statements of "laws" that might serve as appropriate axioms, for example, are hardly to be found. If, on the other hand, one goes to these works looking instead for descriptions of families of models, the task becomes far more tractable.

Here we are concerned with the theory that preceded the modern theory of plate tectonics. For that we must consult works published, say, around 1950 or earlier. An obvious work is Sir Harold Jeffreys's massive treatise, *The Earth* (1924, 5th ed. 1970). This work is hardly representative, however, because Jeffreys was somewhat militant in arguing the cause of geophysics as opposed to mere geology.

The early editions begin with a chapter on the nebular hypothesis of Laplace. This sets the stage for all that follows because the earth is represented as an originally molten sphere that gradually cooled according to well-known laws of physics. The models of how the earth cooled present a simple pattern. Heavier materials gravitated toward the center, forming a core, leaving lighter materials toward the exterior. At the exterior is a relatively thin crust. The crust is coolest, with generally increasing temperatures toward the interior. As heat escapes into space, the interior cools further and therefore contracts, producing deformations in the crust. Large-scale movements in the crust, the result of contraction in the underlying material, are generally in a *radial* direction. In these models no significant forces are directed tangent to the surface of the crust.

One example of just how literally geologists construed contractionist models is a series of experiments described by Walter Bucher in his 1933 treatise entitled *The Deformation of the Earth's Crust* (1933, 115–24). In one set of experiments, performed in the 1920s, inflated rubber balls were covered with a mixture of paraffin and vaseline. They were then slowly deflated under water in a pressure tank. The idea was to see if one could thus mechanically reproduce something like the overall pattern of deformation exhibited by the earth's crust. These experimenters seem to have been fairly pleased with their results.

It is an immediate consequence of these simple contractionist models that, once the major contractions have taken place, the gross features of the earth's surface remain fairly stable. That this consequence was explicit appears in the following statement dating from 1910 by Bailey Willis, a paleontologist and paleogeographer, and professor at Stanford University, who was elected president of the Geological Society of America in 1929. "The great ocean basins," he wrote, "are permanent features of the earth's surface and they have existed where they now are with moderate changes of outline since the waters first gathered" (Willis 1910, 243).

Wegener and Continental Drift

The idea that the continents might have undergone significant lateral displacement has a long history, and it gained much from the gradual mapping of the shorelines of North and South America. The complementary shapes of Africa and South America, in particular, invite speculation that the two were once joined. That speculation, however, was often combined with the biblical tradition of catastrophism—the opening of the Atlantic ocean being associated with Noah's Flood. As such it was far outside the growing scientific tradition in geology, which, after Lyell's *Principles of Geology* (1830), was predominantly *uniformitarian*. By 1900 anyone attempting to make a scientific case for large-scale lateral displacements of the continents was sure to face strong opposition.

ALFRED WEGENER

The person who did the most to bring the idea of laterally drifting continents into the context of genuine scientific debate was Alfred Wegener. Wegener is an exemplar of the innovative scientist who begins at the margins of the field. His doctoral degree, in 1905, was in astronomy, but his dissertation research consisted in transcribing the Alphonsine tables of planetary motions into decimal notation, which was a more historical than scientific topic. He also studied meteorology and atmospheric physics, and his first academic position was in meteorology, practical astronomy, and cosmic physics. His first book (1911) was on the thermodynamics of the atmosphere. One would not then, in 1911, have predicted that he would come to have a major impact on the science of geology.

We know little of Wegener's initial thinking about continental drift. Most accounts simply repeat what Wegener himself said in the "historical introduction" to the fourth edition of his book *The Origin of Continents and Oceans* (1966, 1):

> The first concept of continental drift first came to me as far back as 1910, when considering the map of the world, under the direct impression produced by the congruence of the coastlines on either side of the Atlantic. At first I did not pay attention to the idea be-

cause I regarded it as improbable. In the fall of 1911, I came quite accidentally upon a synoptic report in which I learned for the first time of paleontological evidence for a former land bridge between Brazil and Africa. As a result I undertook a cursory examination of relevant research in the fields of geology and paleontology, and this provided immediately such weighty corroboration that a conviction of the fundamental soundness of the idea took root in my mind.

This is hardly the whole story. Indeed, it is reported that a friend from Wegener's student days, W. Wundt, recalled Wegener as having been impressed by the congruence of coastlines as early as 1903 (Georgi 1962). If that is so, the idea lay fallow in Wegener's brain for a long time before 1911.

A few months later, in January 1912, Wegener gave two lectures expounding the general idea of his theory. In 1914 he was drafted into the military, but he soon was wounded. While recovering, he reworked and expanded his lectures into the first edition of *Die Entstehung der Kontinente und Ozeane* (1915). This is no neutral presentation. Wegener here advocated a displacement theory and marshaled all the evidence he could find in its favor.

The Greenland Experience

Wegener had been fascinated with Greenland from an early age. His dream of going there was fulfilled in 1906, when he was invited to accompany a two-year Danish national expedition as a meteorologist. The experience provided Wegener with a valuable "cognitive resource." He observed glaciers, ice floes, and icebergs. These provided him with valuable models, and a rich source of metaphors, for large masses moving imperceptibly across the surface of the earth. And he used these models, as in the following, typical passage (1924, 2, emphasis added):

> Millions of years ago the South American continental plateau lay directly adjoining the African plateau, even forming with it one large connected mass. This first split in Cretaceous time into two parts, which then, *like floating icebergs, drifted farther and farther apart.*

Many of Wegener's contemporaries found it unimaginable that the continents could move. For Wegener it was easily imaginable.

THE DECISION PROBLEM

There is no way now to determine whether Wegener ever explicitly and consciously *decided* that continental drift was a reality. Yet it is useful to represent him as having at least implicitly done so at some time in the fall of 1911. Doing so helps us to organize his beliefs and interests so as better to understand his actions.

Wegener did not start out with a specific model. What he was choosing was more like what I earlier called an *approach*. It was an approach to representing the history and structure of the surface of the earth that incorporated large-scale lateral movements of the continents. Let us call this approach to the problem a mobilist approach, or, more simply, *mobilism*.

The standard approach was "contractionist," but this term refers more to the dynamics of the earth than to its surface history. A closer parallel to mobilist models is provided by a recognized consequence of standard contractionist models, namely, the laterally stable positions of the continents. The alternative to a mobilist approach, then, is a stabilist approach, or *stabilism*. There were, in fact, many varieties of stabilist "theories" from which to choose.

The corresponding options I shall label simply as "choose mobilism" and "choose stabilism," leaving ambiguous whether the choice is one of "pursuit" or "acceptance." Although this latter distinction is clear enough conceptually, it is often difficult to apply in practice. By his own account Wegener seems to have moved from pursuit to acceptance in a matter of a few months, perhaps only weeks. In any case, the final decision matrix is that shown in figure 8.1.

WEGENER'S INTERESTS

A decision theoretic consequence of Wegener's having been a "marginal" person with respect to geology is that he placed no special value on the

	STABILISM CORRECT	MOBILISM CORRECT
CHOOSE STABILISM		
CHOOSE MOBILISM		

Figure 8.1. The structure of the decision problem faced by Wegener and later by others who seriously considered the possibility of choosing a mobilist approach.

"correct" outcome associated with a stabilist approach. He apparently had no particular geological commitments at all before 1911.

But Wegener was surely aware that his advocacy of a mobilist approach would meet stiff resistance and possibly damage his professional career. Indeed, his later inability to secure a regular professorship in Germany has been attributed to his controversial views (Georgi 1962, 317). For most people that prospect alone would have been sufficient to give the "correct" outcome associated with mobilism a relatively lower value than the corresponding "correct" outcome associated with stabilism. And that might well have biased the choice toward stabilist models. Why was Wegener not deterred, or his advocacy of mobilism at least moderated?

Here I suspect one must appeal to personal variables. Wegener did have at least one influential friend, his future collaborator and father-in-law, Wladimir Koeppen.

WEGENER'S EVIDENCE

Wegener took his evidence from many sources, from geology and geophysics, from paleontology, paleobotany, and paleoclimatology. Here I shall note just two sources because they nicely illustrate how well his reasoning fits a satisficing model.

Already in the first edition of his book Wegener placed great emphasis on the striking geological similarities between several areas in Africa and the areas in South America that would have been adjacent if South America and Africa had once been joined. He noted in particular the detailed geological similarities between the South African Cape mountains and the sierras near Buenos Aires. His own analogy for explaining his reasoning is worth quoting (Wegener 1924, 56):

> It is just as if we put together the pieces of a torn newspaper by their ragged edges, and then ascertained if the lines of print ran evenly across. If they do, obviously there is no course but to conclude that the pieces were once actually attached in this way. If but a single line rendered a control possible, we should have already shown the great possibility of the correctness of our combination. But if we have n rows, then this probability is raised to the nth power.

Earlier on he said that such a match of features "reminds me of the use of a visiting card torn into two for future recognition (1924, 44).

The point of such geological evidence seems to be this. If a mobilist approach is correct, the existence of strong geological similarities at corresponding locations in Africa and South America is very probable. On the other hand, if a static model is correct, the existence of such similarities is highly improbable. These are just the model-based probabilities needed to implement a satisficing strategy. The decision rule is obvious: Choose mobilism if the similarities exist; otherwise, choose a static approach.

The pattern of argument is similar when Wegener comes to consider paleoclimatic arguments for mobilism. By the first decade of this century it was fairly well established that there had been ice caps with considerable glaciation in South Africa, South America, India, and Australia—areas that are now temperate or warmer. Those ice caps had all existed in the late Carboniferous or early Permian periods, roughly 250 million years ago. The standard explanations typically assumed some sort of "polar wandering," that is, that the poles of the earth have shifted their position, or, alternatively, that the crust has shifted as a whole around the core. Wegener argued that hypothesis was inconsistent with the data, particularly when one remembers that both poles must maintain their relative positions to each other. He summed up his argument as follows (1924, 97):

> The more exactly and completely we understand the whole evidence of the climates of those times, the more evident it becomes that, with the present positions of the continents, those climates cannot be adapted at all to any possible position of the poles and the climatic belts.

In other words, according to stabilist models the existing evidence would be highly improbable. The very next paragraph continues (1924, 98):

> The riddle of the Permo-Carboniferous glacial period now finds an extremely impressive solution in the displacement theory: directly those parts of the earth which bear these traces of ice-action are concentrically crowded together around South Africa, then the whole area formerly covered with ice becomes of no greater extent than that of the Pleistocene glaciation on the northern hemisphere. There is no longer merely a question of simplification which the displacement theory provides, it rather affords the first possibility of any explanation whatsoever.

Given that other conditions besides just lateral displacement of continents are required for the formation of an ice cap, Wegener's mobilist model by itself does not make the observed glaciation highly probable. But it does make it possible, maybe plausible, and perhaps at least somewhat probable. In any case, for Wegener this evidence contributed toward making the *overall* decision in favor of mobilism a satisfactory one.

WAS WEGENER RIGHT?

Wegener's conclusion, of course, turned out to be correct. But one might want to ask whether he was right to draw that conclusion when he did. Did the evidence warrant the conclusion? Was his choice a *rational* choice? The reader will by now realize that I raise these questions only to dismiss them. There are no standards of "warrantability" or "rational choice" that could sustain answers to these questions. This is not to say that we cannot go a long way toward *explaining* Wegener's choice. We can even argue that, as an empirical hypothesis, he was following a generally effective strategy. But then so were his critics.

Wegener's Critics

A thin book published in 1915 by an obscure German scientist upholding a scientifically unpopular thesis could not expect widespread attention. But it received enough attention within the German-speaking world to merit a second edition in 1920 and a third in 1922. It was the third edition that attracted widespread interest because it was soon translated into several other languages, including English (Wegener 1924). Here I will focus on a debate that took place in New York City in November 1926 at a "Symposium on the Origin and Movement of Land Masses Both Inter-Continental and Intra-Continental, as Proposed by Alfred Wegener." That symposium, sponsored by the American Association of Petroleum Geologists, featured several leading American figures in the relevant sciences and included Wegener himself. The contributions were published in 1928 under the title *Theory of Continental Drift* (van der Gracht et al. 1928).[2]

There were 14 participants in the symposium. Five, including Wegener, generally supported the drift approach. Four others, while generally critical, thought the drift approach potentially fruitful and at least worth discussing. The remaining five were more uniformly negative, with three of these overtly hostile.

The chairman of the symposium, W. A. J. M. van Waterschoot van der Gracht, provided a summary that began with the following paragraph (1928, 197):

> The criticism which is voiced in this symposium is largely directed against Wegener's conception of continental drift. The arguments can be divided principally into geophysical arguments

about the possibilities and explanation of drift, and geological and paleontological arguments against the facts cited in support of Wegener's drift and the consequences which a drift, such as the exponents of his theory sponsor, would be expected to have.

Philosophers of science have generally taken the notion of an "argument" in the logician's sense, that is, as a set of statements divided into premises and conclusions. They have then sought to understand and evaluate arguments in terms of special "evidential" relationships between premises and conclusions. The approach to be followed here is quite different.

Each of the participants, as well as their colleagues in the more general scientific community, implicitly faced a decision problem having an overall structure like that of Wegener's decision as represented in figure 8.1. In the 1920s the option of "choosing mobilism" typically amounted to the weak decision to regard mobilism as worth pursuing—there might be something to it. The contrary option was to reject a mobilist approach as not worth pursuing and to continue pursuing a stabilist approach. The "arguments" offered are not to be understood in the logicians' sense, but as hypotheses or data that bear on the decision problem in various ways. In fact, van der Gracht's categorization of the "arguments" turns out, by such an analysis, not merely to reflect hypotheses and data from different disciplines. These two classes of arguments bear on the decision problem in very different ways. Moreover, a decision-theoretic analysis explains the almost universal judgment by the participants, as well as by later commentators, that the *geophysical* arguments carried the most weight.

THE ROLE OF GEOPHYSICAL CONSIDERATIONS

The most basic geophysical argument against Wegener's models was that nowhere in or on the earth did there seem to be forces that could account for the lateral movement of continents. Chester R. Longwell, a geologist at Yale University, summarized this argument as follows (van der Gracht 1928, 151):

> The continents in their drifting are supposed to encounter resistance on their forward sides, with the result that the rocks are deformed to make mountain structures, like those of the Andes and the Rockies. If this supposition is correct, the propelling forces not only do the work of transporting the immense continental masses; they do the further enormous work of thrusting and folding the strong continental rocks. . . . The discrepancy between the work performed in making mountains and the forces assumed

for the task is very great. Jeffreys has computed that the forces charged with moving the Americas westward amount to 1/100,000 dyne per square centimeter; the force necessary to make the Rocky Mountains is 1,000,000,000 dynes per square centimeter. . . . The intensity and large scale of mountain deformation imply powerful forces. The puny secular stresses we have discussed appear to be wholly inadequate for the task.

Other geophysical arguments appealed to things like the fact that the rocks underlying the continents are known to be softer than those underlying the oceans. One cannot, it was generally thought, push a softer rock through a harder one.

The conclusion to be drawn from these arguments is that large-scale drift is *physically impossible*. And the grounds for this conclusion lie in widely recognized facts and basic principles of the most secure science, physics. The impact on the decision problem is devastating. If accepted, these conclusions simply eliminate the right-hand side of the matrix in figure 8.1. That is, they eliminate the entire possible state of the world in which a mobilist approach is correct, leaving only the possible state in which a stabilist approach is correct. And that makes the choice of stabilism unavoidable for anyone with the general epistemic goal of being right. This is where all of Wegener's critics started out, and where most quite naturally preferred to remain.

GEOLOGICAL, PALEONTOLOGICAL, AND OTHER CONSIDERATIONS

Earlier I suggested that Wegener's decision in favor of a mobilist approach followed a satisficing strategy. Such a strategy requires several model-based probability judgments. Things that are probable in one approach must be improbable in the other, and vice versa. Now if this was Wegener's strategy, one would expect that many of the objections to his view would take the form of arguing that he was mistaken about various of these probability judgments. And that, indeed, is what we find.

Wegener took it for granted that according to a mobilist model the coastlines of the Americas and Africa would very probably exhibit congruence. Several critics questioned that judgment. Edward Berry, a paleontologist at Johns Hopkins University, put it most succinctly (van der Gracht 1928, 195):

I question altogether the validity of the argument based upon the geographic pattern. It is inconceivable that masses of continental size should move over such large arcs and preserve their outlines of either coast or continental margin intact.

Several critics questioned whether the congruence between the coastlines of the Americas and Europe or Africa is at all remarkable in the first place. The most dramatic presentation in that vein was by Charles Schuchert, professor emeritus of paleontology at Yale University. Schuchert took a globe map of the world, placed a quarter-inch-thick piece of Plasticine over North and South America, and cut it to match the coastlines. He then slid the cutouts over the globe toward Europe and Africa to see how well they fit. The fit was obviously poor. The relevant implication, as I interpret Schuchert's presentation, was that because there is no significant congruence between the coastlines, the existing pattern is not at all improbable within a stabilist approach.

Longwell provided a complementary argument. He produced maps showing that it was possible to produce a fairly good fit of Australia and New Guinea into the Arabian Sea. But no one, Wegener included, believed that Australia had once filled the Arabian Sea. The point of this example, I suggest, is to show that the existence of close geographical fits is not so improbable in a stabilist model.

Wegener also made use of findings from paleontology and paleobotany. The fossil record, he argued, shows striking similarities between earlier fauna and flora on opposite sides of the Atlantic, similarities that would be expected if his model were correct. John Gregory, professor of geology at Glasgow University, challenged the importance of those findings as follows (van der Gracht 1928, 95):

> Arguments based on the biological affinities across the two sides of the Atlantic require that the Pacific Ocean should always have been wider than it is at present. Yet the biological resemblances between southern and tropical America and the opposite coasts of the Pacific are, if allowance be made for the greater width of the Pacific, as striking as those between the two sides of the Atlantic.

The point of this objection, by my analysis, is that even if two continents have not been joined, striking biological resemblances are not all that improbable.

In general the critics who referred to the biological data tended to favor the standard explanation, which appealed to various *land bridges* across the Atlantic, say between Africa and Brazil. If one regards the existence of earlier land bridges as plausible, one can reject a mobilist approach without rendering the known biological similarities highly improbable. But geophysicists had been criticizing the idea of land bridges for a decade or more on grounds similar to those used to criticize Wegener. It is not physically possible for the lighter material of the bridge to sink into the harder material beneath the ocean floor. Nonetheless, several of the participants explicitly expressed confidence that the geophysicists would eventually solve this problem.

The above is but a smattering of the nongeophysical objections raised against Wegener's model. It is enough, however, to show that many of the objections can indeed be understood as attempts to undermine the model-based probability judgments required by a satisficing strategy if that strategy is to lead to a decision in favor of mobilism. The analysis also shows that the chairman was correct in thinking that these latter arguments are importantly different from the geophysical arguments considered earlier.

THE ROLE OF INTERESTS

The above analysis appeals only to the epistemic interest of making a "correct" decision, whichever it might be. But one of the virtues of a decision-theoretic analysis is that it provides a natural place for a variety of other interests as well. If other interests play a role in science, one would expect to see them operative in this case. In fact, a reading of the symposium contributions, together with a modicum of further information about the participants, suggests a very strong positive correlation between opposition to Wegener and easily identifiable professional commitments to stabilism. It would require a very detailed and sensitive study to substantiate, perhaps even to quantify, this impression. Here I will offer only one illustration and a reference to supporting work.

Among the most vehement objectors to Wegener's model was Rollin T. Chamberlin. In a brief five pages he listed eighteen objections to Wegener's model, including several of those noted above. He also referred to Wegener's "dogmatism," "blunders," and "superficial" argumentation. His seventeenth point, in total, reads as follows (van der Gracht 1928, 87):

> Wegener's hypothesis in general is of the foot-loose type, in that it takes considerable liberty with our globe, and is less bound by restrictions or tied down by awkward, ugly facts than most of its rival theories. Its appeal seems to lie in the fact that it plays a game in which there are few restrictive rules and no sharply drawn code of conduct. So a lot of things go easily. But taking the situation as it now is, we must either modify radically most of the present rules of the geological game or else pass the hypothesis by. The best characterization of the hypothesis which I have heard was a remark made at the 1922 meeting of the Geological Society of America at Ann Arbor. It was this: "If we are to believe

Wegener's hypothesis we must forget everything which has been learned in the last 70 years and start all over again."

The appeal to professional interests is right on the surface. No one who has spent a professional lifetime learning the achievements of the past 70 years wants to start all over again.

But there is more. Chamberlin's eighteenth point was that Wegener had attacked only the contractionist models of the Laplacean sort. Chamberlin agreed that those models were inadequate and chided Wegener for not realizing that kind of contraction model had "long been discredited." Chamberlin himself favored the planetesimal hypothesis, which has the earth forming by gravitational attraction from small chunks of matter. The consequent packing and heating of such a body allow for far greater, and less uniform, surface movement than classical contractionist models would. But the hypothesis still implies relatively fixed continents and oceans.

Now, the planetesimal hypothesis was the creation of Rollin Chamberlin's father, Thomas C. Chamberlin, and his colleague at the University of Chicago, astronomer Forest Moulton. Of course the current professor of geology at Chicago, Rollin Chamberlin, would have a very strong preference for the outcome in which stabilism is correct, and he, following in his father's footsteps, wisely chose to pursue it. It would take great confidence in overwhelming probability judgments for such a person to see the choice of a mobilist approach as anywhere near satisfactory.[3]

I have chosen the most extreme case, but several other critics, for example, Bailey Willis, provide a similar story. That professional interests played an important role in the reaction to drift models is also borne out by a recent quantitative study of roughly a hundred geoscientists who published between 1910 and 1950 (Stewart 1986). In this study the most powerful predictor of attitudes toward mobilism was the number of publications—those who published more tended to be less favorable. Since most of the relevant publications over those years were within a stabilist framework, the number of publications is a plausible measure of investment in that approach. In general, it seems to me beyond dispute that professional and other personal interests played a large role in the appraisal of Wegener's model by earth scientists in the 1920s. The only remaining question is how best to evaluate or measure that role.[4]

THE GONDWANA EXPERIENCE

One must not fall into the rationalist trap of thinking that professional and other interests operate only on those who choose what eventually turns out to be the wrong approach. Every scientific decision incorporates a range of interests. Among Wegener's four supporters at the 1926 symposium, one, Frank Taylor, had himself proposed a less well developed mobilist model in 1910. Another, John Joly, had proposed a theory that included less extensive horizontal displacements, but allowed for the possibility of even greater ones. Both, therefore, had obvious reasons to prefer the outcome in which mobilism is correct.

The other two of Wegener's supporters, van der Gracht and Molengraaff, were both from the Dutch school of geologists that had extensive connections with South Africa and the East Indies. This fact is relevant because the geological evidence for some sort of drift is most obvious in the *southern* hemisphere, in the areas that made up the ancient supercontinent of Gondwanaland—South America, Africa, Antarctica, Australia, India, and assorted East Indian islands. Wegener himself laid great stress on the geological similarities of these areas. The relevance of the Gondwana experience is borne out by two recent studies. Working from a bibliography of works on continental drift published between 1900 and 1963 (Kasbeer 1973), Schlosser (1984) identified 311 scientists who published in the earth sciences between 1900 and 1963. Of 69 scientists having some experience with Gondwana materials, 63 (91 percent) supported a drift approach. Of 242 scientists without any Gondwana connection, 116 (48 percent) supported a drift approach. Stewart (1986) reached a similar conclusion from his sample of scientists publishing between 1910 and 1950. The second strongest predictor of his scientists' attitudes toward mobilism next to the number of publications, was having had some experience with Gondwana materials.

Experience with geological materials from Gondwanaland constitutes what I earlier called a *cognitive resource*. Here we have another illustration of a cognitive resource generating scientific interests and thus influencing scientific decisions.

Why There Was No Revolution in the 1920s

There is considerable room for debate about the influence of mobilist models in the earth sciences community following the 1926 meeting and Wegener's tragic death on the Greenland ice cap in 1930. But it is indisputable that nothing like a revolution in the earth sciences took place. Mobilism did not triumph in the 1920s. My explanation for why it did not is simple in outline, though complex in detail. Given the distribution of professional interests and experiences within the relevant scientific communities, there simply were far too few scientists, particularly influential scientists, who could see choosing mobilism as a satisfactory decision. How that decision looked to different individuals varied. For some the geophysical arguments were decisive. Others focused more on the model-based probability judgments required to make mobilism a satisfactory choice. All of Wegener's judgments could be disputed, and most were.

Could things have been different? Of course. Imagine for a moment that the great centers of culture and learning had not been in Europe and North America but in the southern hemisphere, in Argentina, South Africa, Australia, and India. Then most geologists would have had their primary experience with Gondwana rocks. The evidence is that such a difference would have been considerable, perhaps even decisive. Maybe Wegener himself would have been unnecessary.

Does this mean that science is merely a social construct after all? Of course not. It simply means that the decision of a scientific community is a function of the decisions of its members, and that the decisions of individuals are, in part, a function of their individual cognitive resources, some of which are derived from their experiences with the world.

Oceanography and Paleomagnetism

Since Kuhn, theories of the growth of science have presupposed a fairly homogeneous scientific community. But earlier I argued that homogeneity is not true even of the subfield of nuclear physics. I suggested, moreover, that the *variation* in cognitive resources among individual members of a scientific community plays a major role in the *evolution* of the field. Here I should like additionally to suggest that something similar happens at higher levels of scientific organization.

Between Wegener's failed revolution of the 1920s and the eventual revolution of the 1960s, one does not see anything like 40 years of "normal science"—the working out of an established "paradigm" (Kuhn 1962). Nor does one see the steady development of rival "research programs" (Lakatos 1970) or "research traditions" (Laudan 1977). At the individual level one finds varying degrees of involvement with displacement models of the earth's surface. More significantly, following World War II one finds the development of whole subdisciplines of earth science that were relatively insignificant in the 1920s. For the most part those developments initially had nothing to do with the issue of continental drift.

Here I will highlight three subdisciplines that became increasingly prominent following World War II. Two of them turned out to provide the primary means by which the issue of continental drift was finally resolved. The third might have played that role but in fact did not. But all this became clear only in hindsight. With respect to the then minor ongoing debate concerning continental drift, these developments were nothing more than random variations.

OCEANOGRAPHY

The systematic charting of the ocean floor began in the 1870s, but nothing like an extensive map existed until the late 1950s. Among the most striking features of the ocean floor is a vast system of ridges. Beginning in the Arctic the Mid-Atlantic Ridge roughly follows the middle of the Atlantic Ocean all the way to the Antarctic Ocean, where it turns east around the southern end of Africa and then on to the Indian Ocean where it becomes the Mid-Indian Ridge. This system passes between Australia and Antarctica, after which it becomes the Pacific-Antarctic Ridge, which then joins the East Pacific Rise near Easter Island. The East Pacific Rise continues north, apparently interrupted by the San Andreas Fault and others, all the way to Alaska. There are also several offshoots, such as the Carlsberg Ridge in the Indian Ocean and the Chile Rise extending from Easter Island to the tip of South America. Figure 8.2 shows the main outlines of this ridge system as it was conceived around 1960.

As the figure shows, the ridge system is not continuous but is repeatedly offset, sometimes by hundreds of miles. In addition, the Mid-Atlantic Ridge was found to exhibit a wide trough down the middle, in places as large as 30 miles wide, with walls as much as a mile high on either side. The East Pacific Rise has no trough. Both ridges, however, exhibit a higher rate of heat flow along the axis of the ridge than over the ocean floor to either side. Figure 8.3 shows a profile of the Mid-Atlantic Ridge.

In the late 1950s measurements of the *magnetic field* along the ocean floor off the western shore of North America began to exhibit a striking, but very puzzling, pattern. There seemed to be alternating stripes of stronger and weaker magnetism. When the alternating areas were plotted in black and white, the result was the zebra pattern exhibited in figure 8.4. In 1960 there was no accepted explanation of this pattern.

PALEOMAGNETISM

Rocks that form from an originally molten state, like those found in lava flows, exhibit a weak remanent magnetic field aligned in the direction of the earth's magnetic field at the time and place of their formation. In the molten state atoms of magnetic materials, such as iron, are free to line up with the direction of the field. As the surrounding material cools and solidifies, these atoms become locked into position.

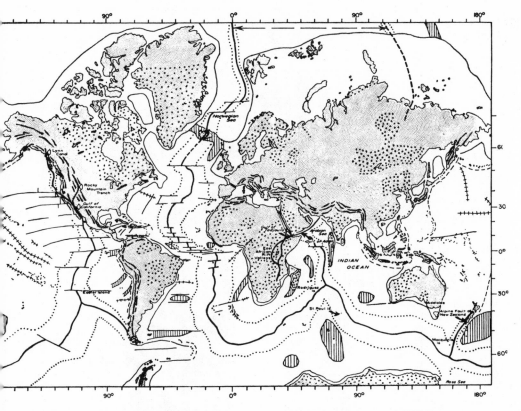

Figure 8.2. The worldwide system of midocean ridges as it was conceived around 1960. Reproduced from Heezen (1962, 260).

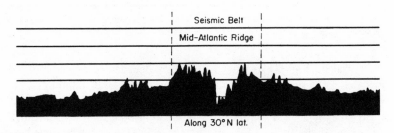

Figure 8.3. A profile of the Mid-Atlantic Ridge. Reproduced from Heezen (1962, 262).

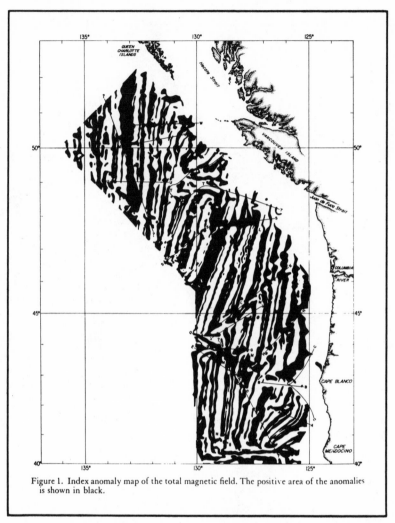

Figure 1. Index anomaly map of the total magnetic field. The positive area of the anomalies is shown in black.

Figure 8.4. An anomaly map of the magnetic field along the ocean floor off the coast of western North America. Reproduced from Raff and Mason (1961, 1268).

Several geologists working in the early decades of this century found volcanic rocks that seemed to have their magnetic fields pointing in the *reverse* direction from "normal" rocks. In 1906 Bruhnes proposed that the earth's magnetic field might earlier have had its polarity reversed from what it is today. In the 1920s the geologist Motonori Matuyama attempted to date samples of reversed polarity and concluded they were of relatively recent origin, roughly 1.5 million years old. He concluded that the earth's magnetic field had reversed its polarity within the last million years or so.

The topic of geomagnetic reversals was revived in the early 1950s (Hospers 1951). In the mid-1950s several young geologists trained at Berkeley took up the problem. By 1963 they had confirmed Matuyama's conclusion and published the first scale of reversals shown at the top of figure 8.5 (Cox, Doell, and Dalrymple 1963). This scale shows the present epoch of

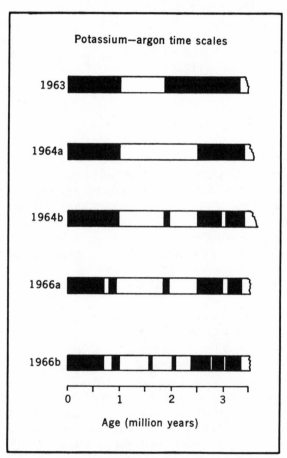

Figure 8.5. The gradual refinement of time scales for geomagnetic reversals based on potassium-argon dating techniques applied to terrestrial rocks. "Normal" magnetism appears in black. Reproduced from Cox (1969, 239), copyright 1969 by the AAAS.

normal magnetism (named "Bruhnes") extending back roughly one million years, preceded by an epoch of reversed polarity of roughly three-quarters of a million years (named "Matuyama"), which in turn was preceded by another normal epoch (named "Gauss") of unknown duration.

During the next several years Cox and his associates refined the scale in both its detail and its accuracy. Of particular interest was the discovery of relatively brief periods of changed magnetism. The first of these so-called events, discovered in rocks from the Olduvai Gorge, was a brief period of normal magnetism roughly two million years ago that lasted only about 100,000 years. This finding led to the scale labeled 1964b in figure 8.5 (Cox, Doell, and Dalrymple 1964). By 1966 these researchers had identified several additional such events. The "Jaramillo" event (Doell and Dalrymple 1966), of about one million years ago, turned out to be particularly significant, as we shall see.

DIRECTIONAL GEOMAGNETISM

About 1950 the British physicist P. M. S. Blackett proposed that any rotating body produces a weak magnetic field. He then attempted to measure the suspected effect. The results of his experiments were inconclusive, but in the process he created an extremely sensitive magnetometer.

One of Blackett's assistants at the time, Keith Runcorn, was interested in investigating the possibility of long-term variation in the direction of the earth's magnetic field, and the new magnetometer proved suitable for measuring the direction of the remanent magnetic field in rocks of various ages. The direction of the magnetism in a rock implies a position for the magnetic pole of the earth at the time the rock was formed—assuming, of course, that the rock has not moved relative to its immediate surroundings in the intervening period.

Runcorn began measuring the direction of magnetism in British red sandstone. He soon discovered that the direction of the magnetic field of rocks of different ages pointed in notably different directions! By 1954 he was confident that the magnetic pole of the earth had been moving, relative to the British islands, over the last several hundred million years.

There are three possible types of explanation for this finding: (1) The British Isles are fixed in place and the magnetic pole has moved. (2) The magnetic pole is fixed and the British Isles have moved relative to the surface of the earth. (3) The pole is fixed, but the whole crust of the earth has shifted. Combinations of these are also possible.

Runcorn and his co-workers, particular Edward Irving and Kenneth Creer, set out to measure the remanent magnet field of rocks from several

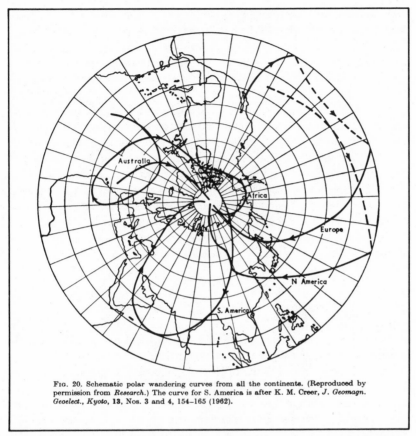

FIG. 20. Schematic polar wandering curves from all the continents. (Reproduced by permission from *Research.*) The curve for S. America is after K. M. Creer, *J. Geomagn. Geoelect., Kyoto,* **13**, Nos. 3 and 4, 154–165 (1962).

Figure 8.6. The apparent position of the magnetic north pole for several continents as a function of time. The arrows indicate time moving toward the present. Reproduced from Runcorn (1962, 24).

continents. By 1960 they had added data from North America (Runcorn), South America (Creer), India (Irving), and Australia (Irving). Figure 8.6 shows Runcorn's plot of the position of the magnetic pole for several continents as a function of time (the arrows point forward in time). Note in particular the curves for North America and Europe. All the most recent rocks point to the current magnetic north pole. But the older rocks indicate an apparent divergence of the pole for Europe and North America. For rocks 200 million years old the divergence is roughly 30 degrees, which is the current angular distance across the Atlantic. Runcorn and his associates drew the obvious conclusion. Wegener was right. Europe and North America have moved apart by 30 degrees in the past 200 million years (Runcorn 1962).

As a matter of fact, these results by the Runcorn group played only a small part in convincing the wider earth sciences community that continental drift is a reality. Looking at their work with 20 years' hindsight, one wonders why they were not more influential at the time. My tentative suggestion appeals to differences in the cognitive resources, as well as the social organization, of other earth scientists. The majority, I suspect, either did not learn about the findings, did not fully understand them, or did not trust them. Blackett and his associates were primarily physicists, not geologists. They were working on a novel problem with novel instrumentation. Few other scientists, including most geologists, possessed the cognitive resources to deal with these issues. I suspect that in time this work would have come to be better understood and appreciated on its own. But events overtook it, and it was simply absorbed in the subsequent revolution.[5]

THE ROLE OF NEW TECHNOLOGY

Earlier I drew attention to the importance of new technology in the progress of modern science. That point is well illustrated by the progress of the earth sciences in the years since World War II. Blackett's magnetometer is a prime example, but it is not an exceptional case.

The discovery and mapping of the magnetic stripes on the ocean floor, for example, were greatly facilitated by the development of the proton precession magnetometer. The original development of this means of measuring magnetic fields had no connection with oceanography, but it proved highly adaptive in that environment. In particular, it could be operated by remote control in a torpedo-like container dragged behind and below a research vessel as it passed above regions of the ocean floor to be surveyed (Glen 1982, 288–92).

Similarly, the determination in the late 1950s of the time and duration of earlier magnetic reversals was originally made possible by the development of a technique for employing potassium-argon dating of minerals—a technique requiring a special application of a mass spectrometer. The problem was that the existing techniques could not, with sufficient accuracy, date rocks considerably younger than four million years old. The fact that this new technique was developed at Berkeley virtually determined that the investigation of geomagnetic reversals would begin there. But the technique was not developed in order to date young rocks. That was an adaptation recognized only later by others on the scene (Glen 1982, 27–49).

If one takes seriously the idea that the progress of science is (at various levels of organization) partly driven by a process of random variation and selective retention (Campbell 1960), one factor whose variation and selection must be considered is the *technology* employed in the gathering and processing of new information.[6]

Seafloor Spreading

As noted earlier, the most persuasive ground for rejecting mobilism was the geophysical argument that any large-scale lateral movement of the continents is physically impossible. Given the established resistance of the ocean floors, there simply are no known forces that could accomplish such a feat. For anyone otherwise inclined against mobilism, for whatever reasons, this geophysical argument made the decision in favor of stabilism trivially easy.

Yet already in the late 1920s Arthur Holmes, later Regius Professor of Geology and Mineralogy at Edinburgh, had suggested an answer (Holmes 1929). Holmes, who pioneered the use of radiometric techniques for dating geological periods, was much impressed with the importance of radioactive heating within the earth (Frankel 1978). He concluded that the effects of cooling and contraction must be much smaller than were supposed by a standard contractionist approach. In fact, he pushed the idea of radioactive heating even further. The heating, he suggested, might be sufficient to create a system of convection currents just below the crust of the earth. Those currents, rising and then spreading out below a land mass, could simply rip it apart and move the pieces laterally, the separation being filled by water. At some point the far edge of the moving segment must be drawn down into the earth to complete the circuit.

The last chapter of Holmes's influential textbook of physical geology is entitled "Continental Drift." In the last section of that chapter, entitled "The Search for a Mechanism," he summarized his suggestion as follows (1944, 508):

> During large-scale convective circulation the basaltic layer becomes a kind of endless travelling belt on the top of which a continent can be carried along, until it comes to rest (relative to the belt) when its advancing front reaches the place where the belt turns downwards and disappears into the earth.

Figure 8.7 reproduces the diagrams that accompany this section.

Even Jeffreys admitted that Holmes's model of convection currents was

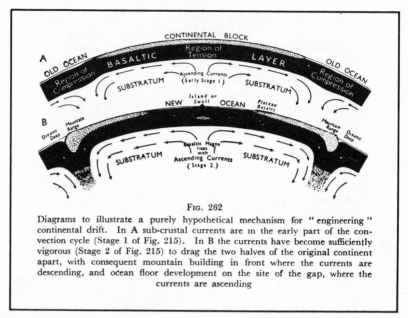

FIG. 262

Diagrams to illustrate a purely hypothetical mechanism for "engineering" continental drift. In A sub-crustal currents are in the early part of the convection cycle (Stage 1 of Fig. 215). In B the currents have become sufficiently vigorous (Stage 2 of Fig. 215) to drag the two halves of the original continent apart, with consequent mountain building in front where the currents are descending, and ocean floor development on the site of the gap, where the currents are ascending

Figure 8.7. Holmes's model of convection currents splitting a continent and producing a new ocean. Reproduced from Holmes (1944, 506).

physically possible. "So far as I can see," he wrote, "there is nothing inherently impossible in it" (Jeffreys 1931, 453). Yet even this admission had little immediate impact on the debate. Wegener, in the 1929 edition of his book, wrote approvingly of convection currents, though without citing Holmes, and without himself embracing the idea (Wegener 1966, 178). And although Du Toit, the chief advocate of mobilism after Wegener's death, invoked Holmes's model in his final chapter on the possible causes of continental drift, he too did not fully embrace it (Du Toit 1937, 321–26). The idea was not to gain currency until 30 years later!

It is worth noting, in passing, that the fate of Holmes's suggestion itself raises a problem for accounts of scientific progress based on notions of "problem-solving effectiveness" (Laudan 1977; Frankel 1980a, 1980b). The problem of finding an adequate mechanism for continental drift is a perfect example of what Laudan (1977, 50–54) called a "conceptual problem." Its importance was widely recognized. Yet a recognized conceptual solution had almost no impact on the subsequent history of the subject. I would suggest that this reflects a simple realism among scientists. They are, in general, not going to decide that a model is right, or even worth pursuing, without some basis for that decision beyond its mere physical possibility. This attitude was expressed by Holmes himself. The very last paragraph of his *Principles of Physical Geology* (1944, 508–9) begins:

> It must be clearly realized, however, that purely speculative ideas of this kind, specially invented to match the requirements, can have no scientific value until they acquire support from independent evidence. The detailed complexity of convection systems, and the endless variety of their interactions and kaleidoscopic transformations, are so incalculable that many generations of work, geological, experimental, and mathematical, may well be necessary before the hypothesis can be adequately tested.

He was just a bit too pessimistic. It took only until 1966.

HESS'S MODEL OF SEAFLOOR SPREADING

In 1960 Harry Hess, professor of geology at Princeton University, revived the idea of convection currents. I am not prepared to trace the development of convection models in Hess's repertoire of cognitive resources, acquired over a long career beginning in the early 1930s. It is clear that he had been familiar with these models for at least 20 years through his association with David Griggs (1939) and Felix Vening Meinesz—the latter being famous for undersea gravity surveys in the Dutch East Indies (Vening Meinesz 1930). Nevertheless, if Frankel (1980a) is right, Hess was opposed to mobilism as late as 1959.

Hess's model of seafloor spreading strongly resembled Holmes's, as is obvious from comparing the diagram Hess offered (Figure 8.8) with that of Holmes (Figure 8.7). And Hess did refer to Holmes. But there is one major difference. The prominence of the oceanic ridges as features of the earth's surface needing explanation became apparent only in the mid-1950s. Hess's model had those ridges being created by rising convection currents directly beneath the ridges, not under a land mass, as in Holmes's model. The consequences of the two models for mobilism, however, are similar. According to Hess (1962, 608–9):

> The Mid-Atlantic Ridge is median because the convection areas on each side of it have moved away from it at the same rate—1 cm/yr. This is not exactly the same as continental drift. The continents do not plow through oceanic crust impelled by unknown forces; rather they ride passively on mantle material as it comes to the surface at the crest of the ridge and then moves laterally away from it.

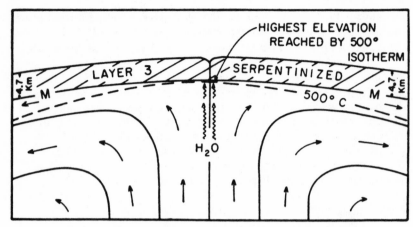

Figure 8.8. Hess's model of convection currents and seafloor spreading. Reproduced from Hess (1962, 607).

This sounds much like Holmes's metaphor of a "travelling belt."

Hess (1962, 107) admitted that "mantle convection is considered a radical hypothesis not widely accepted by geologists and geophysicists." In defense he argued that "if it were accepted, . . . whole realms of previously unrelated facts fall into a regular pattern. . . ." This sounds like an appeal to "unification" or even "problem-solving effectiveness." Yet Hess surely did not regard these "arguments" as sufficient grounds for deciding the model was right. Borrowing a term from Umbgrove (1947), he began his discussion by referring to it as "an essay in geopoetry." And he concluded (p. 618) in much the same vein as Holmes did:

> In this chapter the writer has attempted to invent an evolution for ocean basins. It is hardly likely that all of the numerous assumptions made are correct. Nevertheless it appears to be a useful framework for testing various and sundry groups of hypotheses relating to the oceans. It is hoped that the framework with necessary patching and repair may eventually form the basis for a new and sounder structure.

Unlike Holmes, he had not long to wait.

The Vine-Matthews Hypothesis

In January 1962 Harry Hess was a guest speaker at an interuniversity congress held at Cambridge University. The topic of the congress was "The

Evolution of the North Atlantic," and Hess presented a version of his (at that point still unpublished) paper on the history of ocean basins. In the audience was a last-term undergraduate student majoring in geology, Fred Vine. Few other people could have been more receptive to Hess's model, and, as it turned out, few were as well located to pursue it.

Vine, like Wegener and many others before him, claims he had been intrigued by the similarity in the coastlines around the Atlantic ever since he first saw it while still a schoolboy. He relates (Frankel 1982, 12):

> It seemed rather basic to me, rather fundamental. From that point on, I decided it had to be true; it was too simple and elegant. So I was always very favorably disposed. I guess, I was always looking in studying geology for evidence that confirmed my prejudice.

On another occasion Vine remarked that he "believed in spreading, or wanted to believe in spreading" (Glen 1982, 278). In the context these remarks sound like confessions that he was not properly "scientific" about the topic. My interpretation is just the reverse. The existence of such "prejudices" as Vine's is the normal state of affairs in science and contributes to its overall progress.

In the fall after the January 1962 congress Vine returned to Cambridge as a graduate student in the department of geodesy and geophysics. In addition to his prejudices and Hess's model of seafloor spreading, he had one other powerful cognitive resource: a good grounding in mathematics and physics, his original major fields of study. The switch to geology had come late in his undergraduate career.

Vine was more or less assigned as a research student to Drummond Matthews. Matthews had completed his undergraduate degree at Cambridge in 1955, whereupon he joined an expedition doing geological mapping in the Antarctic. That expedition was based in the Falkland Islands. One of the people on the expedition was Raymond Adie, a student of the South African geologist Lester King—a major supporter of mobilism. Adie also supported mobilism and had written on the geological history of the Falklands from this perspective (Adie 1952). As noted in a conversation Matthews recalled having had with Adie (Frankel 1982, 13):

> [Adie] said, "Oh well, . . . if you don't believe in continental drift just take out a tape measure and measure the Devonian sections in the Falkland Islands." I did, and they were very much impressively the same as the description [given by Du Toit for South Africa]. . . . Inch for inch they measured up. So, I came back quite enthusiastic.

After one false start Matthews engaged in dissertation research analyzing rocks dredged from the floor of the eastern north Atlantic. One chapter of his dissertation dealt with the magnetic properties of those rocks, particularly their *remanent magnetism.*

Upon completing his degree in 1961, Matthews was appointed to the faculty of geodesy and geophysics at Cambridge and given the job of directing the British contribution to the Indian Ocean expedition. He decided to focus on a relatively small section across the crest of the Carlsberg Ridge. He later reported being "absolutely persuaded that the mid-ocean ridge was to be understood in terms of fissure eruptions, and he knew basalts of the seafloor were remanently magnetized" (Glen 1982, 273). He returned to Cambridge in November 1962 with the most detailed magnetic survey of a ridge section then existing, but he "had no idea of connecting sea-floor spreading and reversals" (Glen 1982, 274).

Discovering his new research student, Matthews assigned Vine the task of interpreting the magnetic data from the Carlsberg Ridge. Matthews was of the opinion that they should proceed by building physical models of the ridge. But the senior geophysicist at Cambridge, Sir Edward ("Teddy") Bullard, convinced Vine that the task was better approached using a digital computer, which was then a relatively new technology. Bullard, in fact, was using a computer to derive a best fit of the continents of Africa and South America (Bullard, Everett, and Smith 1965), one that turned out to be far better than Wegener's! And so Vine, drawing on his mathematical abilities, acquired yet another valuable cognitive resource. The programs he used to analyze the magnetic survey data, however, were mainly borrowed from others (Glen 1982, 277–78).

Early in 1963, while Matthews was away on his honeymoon, Vine drafted the paper containing what was to become known as the Vine-Matthews hypothesis. Figure 8.9 reproduces one of the published magnetic profiles. The solid line is Vine's computer analysis of the data. The commentary in the paper (Vine and Matthews 1963, 948) reads:

> Computed profiles . . . assuming . . . uniform normal magnetization bear little resemblance to the observed profiles [the dashed line in figure 8.9]. These results suggest that whole blocks of the survey area might be reversely magnetized. The dotted curve in [figure 8.9] was computed for a model in which the main crustal layer and overlying volcanic terrain were divided into blocks about 20 km wide, alternately normally and reversely magnetized.

This model is represented by the suggestive picture reproduced as figure 8.10, where the crustal blocks normally magnetized are marked *N:*

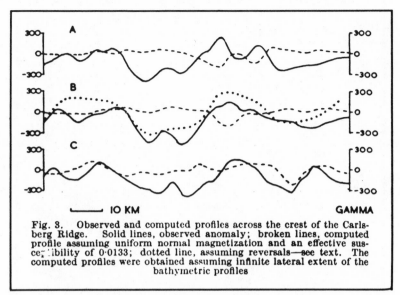

Fig. 8.   Observed and computed profiles across the crest of the Carlsberg Ridge.   Solid lines, observed anomaly; broken lines, computed profile assuming uniform normal magnetization and an effective susceptibility of 0·0133; dotted line, assuming reversals—see text. The computed profiles were obtained assuming infinite lateral extent of the bathymetric profiles

Figure 8.9. Vine and Matthews's comparison of the magnetic data (solid line) across the Carlsberg Ridge, with a model assuming uniform magnetic material (dashed line) and a model assuming alternating magnetism (dotted line). Reprinted from *Nature,* vol. 199, pp. 947–49. Copyright © 1963, Macmillan Journals Limited.

The text of the published paper continues:

> Work on this survey led us to suggest that some 50 percent of the oceanic crust might be reversely magnetized and this in turn has suggested a new model to account for the pattern of magnetic anomalies over the ridges.
>
> The theory is consistent with, in fact virtually a corollary of, current ideas on ocean floor spreading (Dietz 1961) and periodic reversals in the Earth's magnetic field (Cox et al. 1963). If the main crustal layer . . . of the oceanic crust is formed over a convective up-current in the mantle at the centre of an oceanic ridge, it will be magnetized in the current direction of the Earth's field. . . . Thus, if spreading of the ocean floor occurs, blocks of alternately normal and reversely magnetized material would drift away from the centre of the ridge and parallel to the crest of it.

The simplicity of the model itself is obvious even to one with little training in geology. That it might be correct was obvious to almost no one.[7]

The paper concludes with a brief reference to other models (Vine and Matthews 1963, 949):

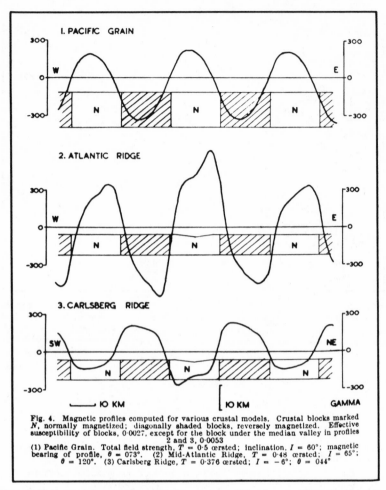

Fig. 4. Magnetic profiles computed for various crustal models. Crustal blocks marked N, normally magnetized; diagonally shaded blocks, reversely magnetized. Effective susceptibility of blocks, 0·0027, except for the block under the median valley in profiles 2 and 3, 0·0053
(1) Pacific Grain. Total field strength, $T = 0·5$ œrsted; inclination, $I = 60°$; magnetic bearing of profile, $\theta = 073°$. (2) Mid-Atlantic Ridge, $T = 0·48$ œrsted; $I = 65°$; $\theta = 120°$. (3) Carlsberg Ridge, $T = 0·376$ œrsted; $I = -6°$; $\theta = 044°$

Figure 8.10 Computed magnetic profiles for several models of ocean ridges. Reprinted from *Nature*, vol. 199, pp. 947–49. Copyright © 1963, Macmillan Journals Limited.

It is appreciated that magnetic contrasts within the oceanic crust can be explained without postulating reversals of the Earth's magnetic field; for example, the crust might contain blocks of very strongly magnetized material adjacent to blocks of material weakly magnetized in the same direction. However, the model suggested in this article seems to be more plausible because high susceptibility contrasts between adjacent blocks can be explained without recourse to major inhomogeneities of rock type within the main crustal layer or to unusually strongly magnetized rocks.

An even stronger view regarding alternative models comes out in Vine's later recollections. "If you worked very closely with the data and the essential problem that was assigned to me as a graduate student," he reported, "then I think you'd realize that almost out of desperation there was no other way of interpreting them" (Glen 1982, 278).

For Vine, at least, the decision to pursue the Vine-Matthews hypothesis fits a satisficing model very well. There is the initial "prejudice" in favor of drift models and a desire that seafloor spreading be correct. There is a simple and direct connection between the model and the observed magnetic data. And there is the judgment that the data would be improbable in any other model.

I shall not try to elaborate the case for Matthews or for others who constructed similar models at about the same time. Later I shall examine the decisions of some of the many critics.[8]

The Juan de Fuca Ridge

There was little immediate reaction to the Vine-Matthews paper, and what there was was predominately negative. Vine's impression was that "it went over like a lead balloon; in some ways there was no response. People just sort of turned away" (Glen 1982, 281). This judgment is borne out by the published record. In the spring and summer of 1964 two international symposia were held on the subject of continental drift, one in Canada (Garland 1966) and one in England (Blackett, Bullard, and Runcorn 1965). Interest in continental drift was definitely picking up. But in the published versions of the symposia only one reference to the Vine-Matthews hypothesis appears (Vacquier 1965), and that is critical in nature.

The next major development further illustrates the "random," evolutionary nature of science. Harry Hess and J. Tuzo Wilson, a senior geological theorist from the University of Toronto, went to Cambridge for the spring term in 1965. Vine, who was just finishing up his dissertation, could have had little to do with initiating their visits. Nor could he have foreseen their impact on his own work. What made the difference for Vine, apart from his considerable cognitive resources, was the additional material resource of being in a place that attracted people like Hess and Wilson.

Wilson arrived in January and shortly took off with his family for a yacht trip along the coast of Turkey. He came back in February bursting with new ideas. One was that the Vine-Matthews hypothesis implied that the pattern of magnetic reversals should be *symmetrical* on either side of the ridge. In Hess's model the seafloor spreads out on both sides of the ridge. Though that was obvious in retrospect, Vine and Matthews had not noted this aspect of their own hypothesis. As Vine later put it, "We simply had not believed in the hypothesis well enough to twig" (Glen 1982, 305).

Another of Wilson's new ideas was *transform faults*—a concept crucial to the later development of what became known as plate tectonics. Hess's model had two sorts of boundaries between moving parts of the earth's surface: ridges where surface material is added, and trenches where it eventually returns into the mantle. Wilson introduced a third kind of boundary, one that exists between adjacent, but laterally displaced, sections of a ridge. These are transform faults.

Figure 8.11 reproduces Wilson's (1965) sketch of the San Andreas Fault as a long transform fault. According to this reconstruction there should be a ridge off the coast of British Columbia, and that ridge should once have been continuous with the East Pacific Rise off the coast of Mexico, a thousand miles to the south. By happy coincidence Wilson's hypothesized ridge lay right in the middle of an area where Raff and Mason had done a detailed magnetic survey several years earlier (Raff and Mason 1961). As Vine later reconstructed the situation (Glen 1982, 306):

> There was one morning when we were all together and Tuzo said, "Look, there should be a ridge here," and Harry Hess said, "Well, if you're going to put a ridge there, then there ought to be some magnetic expression of it on the Raff and Mason map." He pointed out that it was one place where we had good magnetic data, so shouldn't one apply the Vine-Matthews hypothesis if you've got a ridge and you've got magnetics? I hustled up to the library and got out the Raff and Mason map, and lo and behold, for the first time, although it had been in print for four years, we realized that there was symmetry in the anomalies.

As shown back in figure 8.4, Wilson's new ridge, which he and Vine named the Juan de Fuca (after the nearby strait), appears as a broad, black stripe in the center running slightly northeast to southwest. The symmetry to either side is fairly obvious once one begins to look for it.

Figure 8.12 reproduces a diagram from Vine and Wilson's paper published later that year (1965). Here they compared a magnetic profile across the Juan de Fuca with one across the East Pacific Rise far to the south. Both are compared with a simple model like the original Vine-Matthews model. The comparison was, of course, intended to suggest that these now widely separated ridges were produced by the same process.

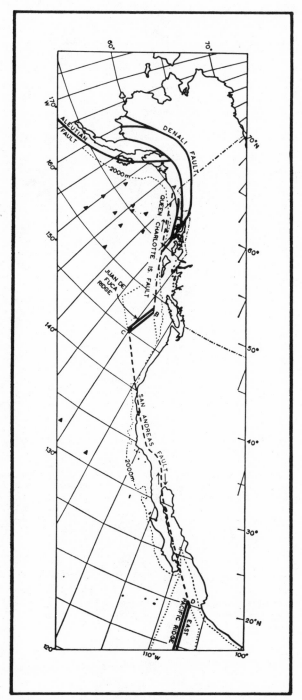

Figure 8.11. Wilson's reconstruction of the San Andreas Fault as a long transform fault. Reprinted from *Nature,* vol. 207, pp. 343–47. Copyright © 1965, Macmillan Journals Limited.

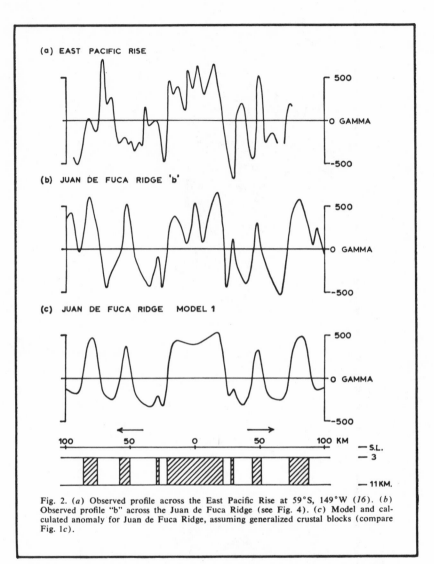

Fig. 2. (*a*) Observed profile across the East Pacific Rise at 59°S, 149°W (*16*). (*b*) Observed profile "b" across the Juan de Fuca Ridge (see Fig. 4). (*c*) Model and calculated anomaly for Juan de Fuca Ridge, assuming generalized crustal blocks (compare Fig. 1c).

Figure 8.12. Vine and Wilson's comparison of the magnetic profiles of the East Pacific Rise and the Juan de Fuca Ridge. The lowest curve is the profile computed from the model at the bottom. Reproduced from Vine and Wilson (1965, 487), copyright 1965 by the AAAS.

A major innovation in the Vine and Wilson paper was that the *widths* of the postulated blocks of alternately magnetized material were no longer the same, but quite variable. The authors drew this conclusion because they had seen the Cox group's 1964 time scale for magnetic reversals (Cox, Doell, and Dalrymple 1964). That scale, shown back in figure 8.5, included the Olduvai event, the brief period of normal magnetization around 1.9 million years ago.

Unfortunately, their identification of the narrow band of normal magnetization in the observed profiles with the Olduvai event yielded a "rather irregular" rate of spreading for the seafloor. The average rate of spreading, however, was about what one would expect if seafloor spreading was a reality. But the main difficulty with the paper was that, like the original Vine-Matthews paper, it depended on a novel idea, in this case Wilson's model of transform faults.

The Vindication of Seafloor Spreading

It is clear that there was some sympathy for mobilism in England, particularly in Cambridge. The opposite was true in the United States. This was definitely the case at the Lamont Geological Observatory of Columbia University. Neil Opdyke, who had been an undergraduate student at Columbia in the 1950s, recalled (Glen 1982, 312–13):

> I came from a department of geology at Columbia University that was dead set against continental drift. Marshall Kay, Walter Bucher, Joe Worzel, and those boys wouldn't hear of it; it was anathema. . . . Ewing was an avid anti-drifter and the leader against mobility, both intellectually and emotionally. Jim Heirtzler and Xavier Le Pichon were similarly disposed.

As Ellen Herron, a graduate student from the period, summed it up, "Doc Ewing's philosophy that the oceans were permanent features was the party line at Lamont" (Glen 1982, 313).

Maurice Ewing was the founder, in 1945, and after that the driving force behind Lamont. He had built it into one of the world's leading centers for research in marine geophysics. He was also, in the 1950s, one of the people most responsible for the discovery that the oceanic ridges constituted a connected, worldwide system. It is an easy speculation that he regarded that system as intellectually his property. If anyone was going to theorize about its true signficance, it was going to be him, or at least someone with whom he could wholeheartedly agree. But that is speculation. It is a matter of published record that in 1965–66 a number of papers coming out of

Lamont included strong criticisms of seafloor spreading and of the Vine-Matthews hypotheses. Lamont was one major center where Vine and Matthews were not ignored.

One paper (Heirtzler and Le Pichon 1965) focused on an analysis of 58 magnetic profiles across the Mid-Atlantic Ridge. The single paragraph devoted to the Vine-Matthews hypothesis concluded (p. 4028), "It is clear from this study that most of the profiles do not follow the pattern assumed by Vine and Matthews."

Another paper (Talwani, Le Pichon, and Heirtzler 1965) discussed new data from the East Pacific Rise and the Reykjanes Ridge southwest of Iceland. This paper devoted a short section to pointing out five problems with the "displacement hypothesis." It devoted three times as much space to elaborating an "alternative hypothesis" according to which "the important motion on the fracture zones, due to the formation of the rise, would then be vertical and not horizontal" (p. 1114). Two paragraphs later they emphasize the importance of Ewing's mid-ocean ridge system in their thinking:

> Basic to our thinking is the view that the major phenomenon we are dealing with is the mid-ocean ridge system, which comes into existence because of changes in the underlying mantle. Preexisting local variations in the properties of the crust and upper mantle decide the exact location the ridge crest, and thus the vertical displacements on the fracture zones are nothing more than important details in the topographic expression of the mid-ocean ridge system.

A third paper (Heirtzler, Le Pichon, and Baron 1966) focused on a magnetic survey of the Reykjanes Ridge done in 1963 and computer-reprocessed by Heirtzler. Figure 8.13 exhibits a version of Heirtzler's profile. The similarity with the newly discovered Juan de Fuca Ridge is striking, but word of Vine and Wilson's latest work had not yet reached Lamont. The Lamont group did notice the symmetry in the profile but took it as evidence, not of seafloor spreading, but simply of some control of the magnetic pattern by material in or under the central ridge. They opened a section of the paper devoted to "Geologic Hypotheses" that might explain the magnetic pattern as follows (p. 440):

> While limits can be put on the configuration and magnetization of these bodies, the origin advocated for this general structural pattern depends on the geologic hypothesis adopted. The location of the axial body under the floor of the rift valley in the Mid-Atlantic Ridge suggests that it consists of volcanic material filling a ten-

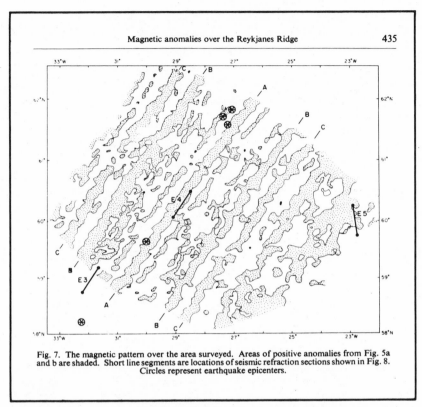

Figure 8.13. The magnetic profile of the Reykjanes Ridge produced by Heirtzler. Reproduced from Heirtzler, Le Pichon, and Baron (1966, 435).

sional crack. The existence of progressively smaller anomalies on the sides was attributed by Heirtzler and Le Pichon (1965) to a pattern of subsidiary fractures. Vine and Matthews (1963), on the other hand, following Dietz (1961), suggested that this pattern was due to a spreading ocean floor, originating at the ridge axis, and alternately normally and reversely magnetized. However, this hypothesis in its present form does not explain the characteristic change in magnetic pattern from the axial zone to the flanks and the difference between the axial anomaly and the adjacent ones.

They did, however, make one important concession. If the magnetic patterns were indeed caused by material in the local environment, the details of the magnetic pattern would be different for different ridges.

My conclusion is that in 1965 it was quite possible for knowledgeable geophysicists to be satisficers and still choose a stabilist approach. One requirement is a high value placed on stabilist outcomes to the decision problem. This was clearly present. The cognitive resources of these researchers (geophysicists rather than geologists), as well as their professional reputations, and perhaps even their personal loyalties, were bound up with stabilist models. Moreover, the models they had were not obviously at variance with the data, as they saw it, and did not render the data improbable in the absence of seafloor spreading and global magnetic reversals. Their only sin was that they turned out to be wrong.

In this case retribution was swift in coming. Before the 1966 paper appeared in print, one of Lamont's own research groups returned with data decisively in favor of Vine and Matthews. And another group then working at Lamont was about to turn up corroborating evidence from yet another quarter. Here the pace of activity became so fast, with several different lines of research coming together at once, that it is difficult to recount in a linear fashion.

THE *Eltanin*-19 MAGNETIC PROFILE

From September through November 1965 one of Heirtzler's graduate students, Walter Pitman, was aboard the research vessel *Eltanin* doing a survey of the Pacific-Antarctic Ridge. Magnetics were not uppermost in their minds. As Ellen Herron, who was also on the cruise, recalled: "We spent most of the time keeping the seismic gear going and didn't know what we were doing with the magnetic data. None of us was aware of the Vine-Matthews hypothesis at the time" (Glen 1982, 332). And Opdyke reported that there was "no mention of the Vine-Matthews hypothesis in the research proposals for those cruises" (Glen 1982, 332).

When they began analyzing the data in December they immediately recognized the similarity between their profiles and those of the Juan de Fuca Ridge just published by Vine and Wilson. But they did not know what to make of it. Pitman recalled (Glen 1982, 333):

> At that point in December, Ellen Herron and I were just getting data out and had not yet selected our doctoral dissertations and didn't know what the *Eltanin* 20 profile really meant. I'd read Dietz's paper casually; I didn't know the Vine-Matthews hypothesis very well at all. I was not aware of it in detail. I had not read the paper and studied it; it was not something that seemed important to me at the time. I was unaware that Le Pichon and Heirtzler had made that strong comment against Vine and Matthews in print.

capability for such study at Lamont. With the increasing interest in paleo-magnetism in the earth sciences community of the early 1960s, Ewing again pushed the idea. In 1964 Lamont hired Neil Opdyke, on funds sup-plied by the Office of Naval Research.

Opdyke, who was from New Jersey, had been an undergraduate student at Columbia, as was noted above. While still at Columbia he served as a field assistant for Keith Runcorn, who was in the United States collecting samples as part of the directional geomagnetic program at Cambridge. Runcorn invited Opdyke to Cambridge for graduate work beginning in 1955. He followed Runcorn to Newcastle-upon-Tyne when Runcorn left Cambridge. Opdyke was thus in the center of the debates over the relative merits of polar wandering and continental drift and wrote his dissertation on "Paleoclimatology and Paleomagnetism in Relation to Polar Wandering and Continental Drift." Upon completing his degree in 1959, he spent a postdoctoral year in Australia, working with Runcorn's associate Edward Irving. By that time most of the directional geomagnetic people had de-cided the drift approach was right, and Opdyke was among them.

Lamont hired Opdyke because he knew about paleomagnetism. He did not know much about deep-sea cores, except that such measurements were both possible and extremely difficult. As sediment forms, any magnetic particles in it naturally line up with the earth's magnetic field. As the sedi-ment is packed down and solidifies, the magnetic particles become locked into place. Thus, if there had indeed been reversals in the earth's magnetic field, different layers of sediment should exhibit magnetism in opposite di-rections. The trouble is that such magnetism would be very weak and diffi-cult to measure. That measurement required instrumentation that, for the most part, simply did not exist. And so despite hints from Ewing, Opdyke was not too keen to embark on the core project, and he attempted to pass it off to his graduate students.

One of those students, John Foster, had considerable skill in electronics and instrument building. In fact, Foster had originally been invited to do Ph.D. work with Heirtzler on the basis of an instrument he had built while working on a master's degree at McGill University. After deciding to study with Opdyke instead, he set to work building a spin magnetometer. But it was not initially designed for work on deep-sea cores. It was designed, in accord with Opdyke's own research interests, and thus Foster's, for use on terrestrial rocks! Moreover, to increase the signal-to-noise ratio of his mag-netometer, Foster designed it with a very low spin rate of only five cycles per second. This rate turned out to be ideal for measuring magnetism in core samples, which, being fairly soft, would lose their magnetic orienta-tion under higher spin rates. But again, this feature of the design was not introduced for that purpose. As Foster himself later put it, "I didn't know a thing about deep-sea cores and couldn't have cared less" (Glen 1982, 329).

The circumstances under which Foster's magnetometer came to be used to measure magnetism in deep-sea cores were equally fortuitous. A graduate student at the University of Rhode Island had heard about Foster's magnetometer and came to borrow the design in order to determine magnetic properties of deep-sea cores for a researcher at the University of Miami. In the course of discussions with this student Foster and several other graduate students at Lamont decided to try some measurements on cores with which one of them was familiar because his office mate had used them for his dissertation research. Foster recalled (Glen 1982, 329):

> Glass took out a segment of core; we carefully marked it and cut out a sample and a meter down the core another sample, and another meter still further down. We shaved the three specimens, preserving their orientation. We measured the first one; [its magnetization] was strong and easily recognizable. Then we did the next one and it was reversed; then we did the next one and it was normal.

They knew immediately they were on to something and called their respective advisors, who included Opdyke and a senior Lamont researcher, Bruce Heezen. Opdyke was upset by the sudden influx of Heezen's group into his paleomagnetics laboratory and the dispute went all the way up to Ewing himself. But the detailed examination of magnetism in the deep-sea cores proceeded full steam ahead.

All this was happening at just the time Pitman was in the very next room analyzing the magnetic profiles from the *Eltanin* cruise. It very soon became clear that both groups were seeing the same pattern! The record of geomagnetic reversals that Pitman saw laid out under the ocean floor was also to be found in the sediments covering that same floor. Figure 8.15 reproduces the results published the following October by Opdyke, Glass, Hays, and Foster (1966). The figure shows the results of their analysis of seven different cores from the south Atlantic. The pattern of reversals among the cores is remarkably similar. They differ primarily only in the rate of sedimentation.

PUTTING IT ALL TOGETHER

Only one more major piece of the puzzle is missing. Partly as a result of his friendship with Hess, Vine was invited to join the faculty of Hess's department at Princeton, beginning in the fall of 1965. In November Vine

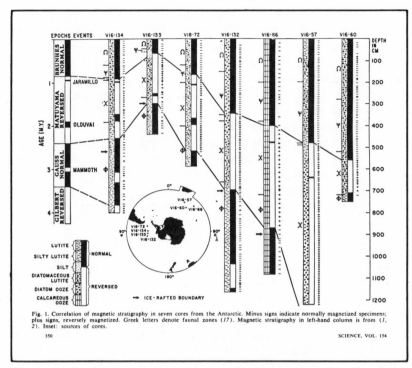

Fig. 1. Correlation of magnetic stratigraphy in seven cores from the Antarctic. Minus signs indicate normally magnetized specimens; plus signs, reversely magnetized. Greek letters denote faunal zones (17). Magnetic stratigraphy in left-hand column is from (1, 2). Inset: sources of cores.

350   SCIENCE, VOL. 154

Figure 8.15. A comparison of magnetic reversals revealed in deep-sea sediments from several locations in the south Atlantic. Reproduced from Opdyke et al. (1966, 350), copyright 1966 by the AAAS.

attended the annual meeting of the Geological Society of America, where he and Wilson presented their papers on the Juan de Fuca Ridge. More important, Vine met Brent Dalrymple. Vine recalled (Glen 1982, 310):

> The crucial thing at that meeting was that I met Brent Dalrymple for the first time. He told me in private discussion between sessions, "We think we've sharpened up the polarity-reversal scale a bit, but in particular, we've defined a new event—the Jaramillo event." I realized immediately that with that new time scale, the Juan de Fuca Ridge could be interpreted in terms of a constant spreading rate. And that was fantastic, because we realized that the record was more clearly written than we had anticipated.

The Jaramillo event is the short period of normal magnetism appearing at around 0.9 million years, as on the 1966 scales in figure 8.5. Because

Vine and Wilson had misidentified the Jaramillo event with the Olduvai, they had concluded that the spreading rate of the seafloor was rather irregular. With the proper identification the spreading rate is amazingly constant. It does, however, differ in different regions of the world. One of the things that makes the *Eltanin*-19 profile so informative is the relatively high spreading rate of the Pacific-Antarctic Ridge, 4.5 centimeters per year. This rate makes even relatively short reversals detectable.

Vine visited Opdyke at Lamont in February 1966. That was in the midst of the Lamont groups' excitement over both Pitman's magnetic profiles and Opdyke's core studies. The Lamont scientists believed they had discovered a new event at 0.9 million years. They had even proposed a name for it. Vine brought the news that the Cox group had already discovered that event and named it after Jaramillo Creek in New Mexico. In return he got to see the *Eltanin*-19 profile for the first time. He, like everyone else, was amazed by its clarity and definition. Even Vine had not thought the record could be that clear.[9]

All these results were first publicly presented to the earth sciences community at the April meeting of the American Geophysical Union. Allan Cox chaired the session. Heirtzler (not Pitman!) presented an analysis of the *Eltanin* profiles, which appeared in print in December (Pitman and Heirtzler 1966). Vine compared the *Eltanin* profile of the East Pacific Rise with those of the Juan de Fuca and the Reykjanes Ridges and drew out the implications for seafloor spreading. Figure 8.16 shows the comparison as it appeared in print that December (Vine 1966). Finally, Opdyke presented an initial look at the evidence from the deep-sea cores. Allan Cox summed up his sentiments about all these findings some years later: "That was the most exciting year of my life, because in 1966, there was just no question any more that the seafloor-spreading idea was right" (Glen 1982, 339).

Mobilism Becomes a Satisfactory Option

Earlier I argued that there was no revolution in the 1920s simply because there were too few earth scientists for whom a mobilist approach seemed a satisfactory option. My explanation for the successful revolution in the 1960s follows the same pattern. It occurred because for most earth scientists, even those with strong contrary interests, choosing mobilism over its stabilist rival had become an obviously satisfactory option. Why was this so? What made the difference?[10]

Here it is helpful to consider the types of arguments offered against seafloor spreading, such as those published by Lamont scientists like Heirtzler and Le Pichon in 1965. One type of argument was to point to features of the

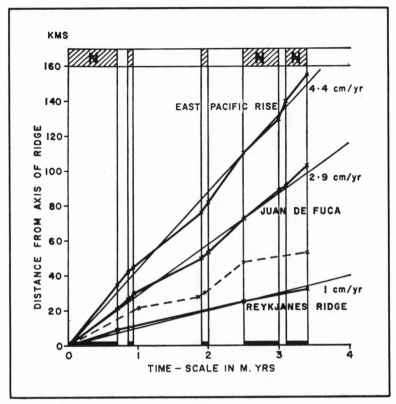

Figure 8.16. Vine's comparison of seafloor spreading rates for three ridges. Reproduced from Vine (1966, 1410), copyright 1966 by the AAAS.

data that were not accounted for by the Vine-Matthews model. That the magnetic anomalies were weaker and more erratic the further they were from the ridge was one such feature. The purpose of such arguments, from a decision-theoretic point of view, was to *eliminate* the Vine-Matthews model as representing a possible state of the world. That left the stabilist approach as the only option.

There was a weakness to those sorts of arguments. At their best they showed only that the particular model suggested by Vine and Matthews did not fit. That it did not is not surprising. The model was very simple, as Vine and Matthews readily admitted. What the arguments failed to show was that no more complex model of the same type could explain the data. This failure illustrates the general point that it may be easy experimentally to reject a *particular model,* but it is very difficult to eliminate a *whole approach.* And this point was even tacitly admitted by the critics themselves when they referred to the Vine-Matthews model with qualifiers like "in its present form." But, of course, they had no interest in exploring other forms of the model. Indeed, they had every interest (in a slightly stronger sense) not to.

The second sort of argument offered by critics of the Vine-Matthews hypothesis was to show how the data cited by Vine and Matthews could be explained using a stabilist model. Understood in terms of a satisficing account of scientific decision making, the purpose of this sort of argument is clear. It is to demonstrate that the data are *not* improbable if the seafloor spreading approach is wrong. This is the crucial model-based probability judgment in any decision between mobilist and stabilist approaches. And this is precisely the judgment that was changed by the *Eltanin* profiles and the magnetic record found in deep-sea cores.

The Lamont critics admitted in their 1965 publications that their model gave no reason to expect that the *detailed* structure of magnetic anomalies across ocean ridges would be the same. Similar in overall pattern, yes, but not in detail, because the pattern would be affected by local conditions. What was found, however, was a remarkable match in the detail of the anomalies across ridges thousands of miles apart, even in different oceans. Moreover, the pattern matched the time scale derived from terrestrial rocks. And both corresponded to the pattern of magnetic reversals found in deep-sea sediments. Not even the most ardent stabilist could imagine any stabilist model in which the existence of these similarities would not be fantastically improbable, or even physically impossible. It was as if nature had left a complex magnetic signature in different forms and at different places around the world. Mobilism offered the only imaginable explanation of how that signature had come to be in those forms at those places.[11]

For even a minimally open-minded satisficer the decision was clear. Choose mobilism. We need not suppose that interests other than the epistemic ceased to operate. Rather, these other interests were simply overwhelmed by the data together with the strong model-based probability judgments just noted. Thus, although difficult, it is possible for experimental evidence to be so strong (in the context of equally strong model-based probability judgments) as to produce the overthrow of a whole approach.

Because the absence of a plausible mechanism for continental drift was so often used to reject Wegener's theory, it is important to emphasize that the 1960s decision in favor of mobilism did not depend on there being a new, detailed model of the mechanisms of continental drift. The mechanism of convection currents suggested by Hess was little different from that proposed by Holmes in 1929. A decision in favor of mobilism included a decision that some type of convection model was correct.

It is sometimes suggested, for example, by Glen (1982), that a major difference between the 1920s and the 1960s is that earlier both the models and the data were mainly qualitative, while later both became much more quantitative. A decision-theoretic analysis explains part of the virtue of more quantitative models. A *quantitative* prediction from an incorrect but quantitative model is much less likely to be confirmed accidentally than a *qualitative* prediction from a similarly mistaken qualitative model. Thus, with quantitative models the model-based probability judgments required by a satisficing strategy are much more secure.

Finally, Frankel (1982), who has sympathies with Laudan's (1977) "problem solving" approach to understanding science, has suggested that the Lamont scientists switched to mobilism in large measure because the problems they had earlier associated with that approach were resolved. It is true that those problems were soon resolved, but I doubt that was why the Lamont scientists changed their minds. In particular, the temporal order seems to be wrong. The resolution of the earlier problems, such as the ambiguity of the magnetic record at greater distances from a ridge, seems to have come *after* most of the original opponents of mobilism had already switched sides.

My view, of course, is that they switched because the new data were wildly improbable in any imaginable stabilist models, but probable in a mobilist model. That made mobilism a satisfactory option. The earlier objections to the Vine-Matthews hypothesis were initially just swept aside. This assessment is borne out by a reading of later works by these same scientists. The following introduction to a much cited article by Heirtzler, Le Pichon, and others (1968, 2119), for example, refers not at all to earlier problems resolved, but to those things that are improbable in any but a mobilist model:

A magnetic anomaly pattern, parallel to and bilaterally symmetric about the mid-oceanic ridge system, exists over extensive regions of the North Pacific, South Pacific, South Atlantic, and Indian oceans. It has been further demonstrated that the pattern is the same in each of these oceanic areas and that the pattern may be simulated in each region by the same sequence of source blocks. . . . The symmetric and parallel nature of the anomaly pattern and the general configuration of the crustal model conform to that predicted by Vine and Matthews (1963) and may be regarded as strong support for the concept of ocean floor spreading (Dietz, 1961; Hess, 1962).

A REALIGNMENT OF INTERESTS

Although Heirtzler may have taken a few weeks to change his mind about drift, and Ewing a few months, that is a short time for so great a change in cognitive orientation. This abruptness suggests there may have been more than just epistemic interests moving the critics toward mobilism.

If the crucial data had been at any other laboratory than Lamont, the Lamont scientists would undoubtedly have spent much more time questioning whether people were really seeing what they claimed to be seeing. After all, as late as 1965 they were still not terribly impressed with the experimental work of Cox and associates on geomagnetic reversals. But these were Lamont data gathered with Lamont instrumentation. No group was better placed to evaluate the data.

Thus, at the same time as the Lamont scientists were realizing that seafloor spreading might be a reality, they were also realizing that the crucial data supporting the new view were in their possession. This gave them a strong, positive interest in seafloor spreading, and they moved quickly to exploit it. During the years 1966–68, a number of major papers developing the mobilist approach came out of Lamont, including that by Heirtzler and others quoted above.

WHAT GETS NEGOTIATED?

Any sociologist of science who happened to be at Lamont in 1966 would have seen many instances of social negotiation. There was, for example, the difficult negotiation between Opdyke and Heezen, mediated by Ewing, over the use of the paleomagnetics laboratory. Later they argued over whose name would appear on which papers, and when.

An even more serious negotiation took place involving Pitman and Vine. Shortly after Vine's visit to Lamont in February 1966, Heirtzler sent Vine the data on the *Eltanin*-19 profile, apparently without telling Pitman. Vine wanted to use the data in a long paper he was preparing (Vine 1966), but Pitman, naturally, did not want to be scooped. Vine rejected Pitman's suggestion that they publish together. Eventually, the editors of *Science* were brought in on the negotiations. The editors wanted to publish the two papers together but agreed to publish Pitman's two weeks before Vine's. As Pitman later put it, "the guys that run *Science* understand geopolitics" (Glen 1982, 337).

'Geopolitics' is exactly right. These were essentially *political* negotiations—who gets what when. There is no suggestion that what was being negotiated was the earth scientists' picture of reality. The outcome of that battle had in effect been decided. Vine and Pitman were merely engaged in dividing up the spoils.

Revolution or Evolution?

The historiography of recent events in the earth sciences has been dominated by Kuhn's theory of scientific revolutions. This is especially true of commentators who are not professional historians of science such as Cox (1973), Hallam (1973), Marvin (1973), or Wilson (1968). Hallam (1973, 108), for example, concluded that "the Earth sciences do indeed appear to have undergone a revolution in the Kuhnian sense. . . ." The reason for the appeal to Kuhn is obvious. When, around 1970, people began reflecting on the great changes that had taken place in the 1960s, Kuhn's work, published in 1962, provided the most visible and most accessible model then available.

Many of Kuhn's critics have objected that change in science is more evolutionary than revolutionary. Often this objection merely indicates that change is more gradual and much less comprehensive than Kuhn seems to imply. And there is much to this minimal objection, even though the changes in the earth sciences were concentrated into a relatively few years. While earth scientists disagreed sharply and were often unsympathetic in their criticism, there is little evidence of anything like "incommensurability," whether understood in terms of the meaning of words or in terms of standards of appraisal. Indeed, people like Heirtzler and Le Pichon were able to move easily from being severe critics to ardent exponents in the space of a year.

My juxtaposition of 'revolution' and 'evolution' is intended to have a deeper significance. In talking of revolution, Kuhn explicitly draws on a *political* analogy, a political revolution being a change in regimes with a period in between that lacks any recognized authority or standards. In invoking the notion of evolution, I, like Toulmin (1972), mean to invoke a *biological* analogy. And not just in the broad sense of gradual change, but in the more specific sense of there being selective mechanisms operating on random variations.

In the previous chapter I argued that the efforts of individual scientists pursuing their own professional interests produced a new theoretical approach to some problems in nuclear physics—and this without any appeal to higher level norms or purposes. At the beginning of this chapter I suggested that the same mechanisms could be seen to operate at more complex levels of organization. Independent developments in the fields of oceanography and paleomagnetics, for example, contributed to the success of a mobilist approach to the earth's history.

Along the way we have seen similar mechanisms operating within a single research organization, Lamont. Individuals in the organization, from Ewing, the director, through senior staff such as Heirtzler, through junior staff such as Opdyke, to graduate students like Pitman, Herron, and Foster, all were pursuing their own professional interests. Yet their interaction produced the evidence that vindicated Hess's model of seafloor spreading—even though that result was irrelevant to or actually contrary to the conscious intentions of most members. Ewing may have been excessive in his adherence to the stabilist approach, but he created an organization that proved remarkably adaptable to a changing scientific environment.

THE COGNITIVE APPROACH

While suggesting that the overall pattern of scientific development is to be seen as an evolutionary process, I would insist that the mechanisms underlying that process are cognitive—not, of course, in the philosophers' sense, but in the sense now common in the cognitive sciences. I have tried to demonstrate the usefulness of this approach in understanding the "arguments" and decisions of earth scientists in both the 1920s and the 1960s, and thus in understanding how changes in the earth sciences did or did not take place. I should like to conclude with some gentle suggestions for all who seek to understand modern science.

Sensitive historians like Frankel and Glen have pointed out important differences among individuals and groups in the earth sciences. They have demonstrated, for example, that British scientists—from the older generation, such Blackett, Bullard, and Holmes, to the younger generation, such as Runcorn and Irving—were generally sympathetic to a mobilist approach. So were the South Africans. The Americans, by contrast, were generally much less sympathetic. Similarly, these historians have noted that geologists and paleobiologists tended to be more sympathetic than geophysicists. It is clear that they note these differences to help explain the beliefs and actions of various individuals or groups. But these historians cannot help conveying a sense of unease about making such observations, perhaps fearing that in citing these differences they are accusing the scientists involved of being less than properly scientific.[12]

Now, if one believed that there are universal, *normative* principles of scientific judgment that all scientists should strive to obey, a sense of unease would indeed be appropriate. Even if one allows that individuals have a less than perfect grasp of the norms, and often have difficulty in applying them, one would expect a far more even distribution of opinion than one in fact finds. The only explanation has seemed to be unscientific bias or other forms of irrationality.[13]

An evolutionary model of science grounded on natural, cognitive mechanisms removes any need to feel apologetic in the face of the obvious fact that the approach to a scientific issue adopted by individual scientists often seems more determined by the accidents of training and experience than by an objective assessment of the available evidence. This is just what one should expect of normal cognitive agents. What sorts of models any individual will regard as most promising or appropriate will of course be strongly influenced by which sorts of models have been learned first and used most. This is not irrationality or anything of the sort. It is normal human behavior, and scientists are normal human beings. Nor does this imply a relativist view of science. The right kinds of interactions among scientists favoring different approaches, together with extensive interactions with nature (mediated by appropriate technology), can produce widespread agreement on the best available approach. That is the lesson of the "revolution" in geology for those who would seek to understand how science works.

Epilogue: Reflexive Reflections

Anyone proposing an empirical model of science cannot help but consider the question of how well that model fares when applied reflexively to itself.[1] Regarding the model presented here, the big question would be: Given the evidence at hand, would many students of the scientific enterprise be inclined to choose a cognitive approach? The answer is all too obvious. The range of interests, both cognitive and otherwise, exhibited by historians, philosophers, and sociologists of science is very wide. The rival models are not articulated well enough to permit strong model-based probability judgments. And the evidence I have presented is itself not that clearcut. In short, there is no reason to predict a revolution in the study of science along the lines of the recent revolution in geology. Indeed, one suspects that no such revolution is possible in any of the human sciences.

On the other hand, it is a feature of my model of science that, as in biological evolution, there is not only one possible path for scientific development. What is produced by the underlying mechanisms of representation and judgment depends greatly on the surrounding environment.

In speculating about the future of relativistic models in nuclear physics, I suggested that relativistic models are likely to win out partly because they agree with the sorts of models that have become standard in other, more fundamental, areas of physics. That sort of development would seem plausible for a cognitive approach to the study of science as well. In the past decade or two the cognitive sciences have developed powerful models of human cognition that now have few competitors. This development provides a strong motivation for bringing the study of science in line with these models. For science, after all, is still one of our best examples of a cognitive activity, complex though it may be. In any case, that scenario now seems to me the most promising one for those of us who favor a cognitive approach.

Notes

1. Toward a Unified Cognitive Theory of Science

1. This is not to say that I fail to appreciate the seriousness of the issue. Philosophers as diverse as Hacking (1983), Nersessian (1984), Nickles (1986), and Rorty (1979) have suggested that there is no "natural kind" for a theory of science to be a theory of. As will emerge shortly, I think there are appropriate natural kinds, namely, cognitive processes, now studied by the cognitive sciences, and evolutionary processes, the subject matter of evolutionary biology and (if this does not beg the question) sociobiology.

2. The general idea of a "cognitive" theory of science, as well as some specific applications, can be found in a number of recent works including Tweney, Doherty, and Mynatt (1981); Tweney (1985); DeMey (1982); Rubinstein, Laughlin, and McManus (1984); Arbib and Hesse (1986); Holland, Holyoak, Nisbett, and Thagard (1986); and Langley, Simon, Bradshaw, and Zytkow (1987). This approach is clearly foreshadowed in such earlier works as Campbell (1959, 1966, 1974); Simon (1957), Gruber (1981), and, of course, in the many works of Jean Piaget (Gruber and Voneche 1977). It can also be found implicitly in Kuhn (1962, 1974). This literature stands in contrast to the older literature in the "psychology of science" (Roe 1952, 1961), which focused on "personality type." To my mind, the best works in this older tradition are Mitroff (1974) and Mahoney (1976).

Many philosophers and philosophers of science have recently turned their attention toward the cognitive sciences with the intent of applying philosophical concepts and methods to investigate these sciences. It is essential to realize that the aim of a cognitive theory of science is just the reverse: to use the concepts and methods of the cognitive sciences to study science itself.

3. A unified view of cognitive science is promoted, for example, by Gardner (1985). This is the best overall introduction to the field currently available. For a somewhat narrower view, more centered on artificial intelligence, see Haugeland (1981, 1985). For a still narrower view see Pylyshyn (1984).

4. The sociologist in question is Andrew Pickering, and the quotation is not

from his very substantial book (1984), but from the discussion following an oral presentation at Indiana University in the fall of 1985.

5. Laudan (1984b) has developed the claim that constructivist sociological accounts fail to explain the success of science.

6. The notion of "schemata" goes back almost a half century to the work of Piaget (Gruber and Voneche 1977) and Bartlett (1932). E. C. Tolman (1948) talked about "cognitive maps" in both animals and humans; for a more recent discussion see Neisser (1976, chap. 6). Johnson-Laird (1983) has popularized the term 'mental model' within cognitive psychology, while Minsky (1975) introduced the term 'frame' into the computer science literature.

7. The combination of anti-realism regarding theories and naturalism regarding scientific judgment matches pretty well with what most people call relativism. In the literature the contrasts tend to be dichotomous, as in Hollis and Lukes's *Rationality and Relativism* (1982) and Laudan's (1984b) discussion of realism and relativism. Failure to sort out the two dimensions, representation and judgment, leads to confusion, as when Laudan suggests that realism and relativism may be "orthogonal," but then pronounces them genuine "contraries."

8. For a clear expression of this point of view in the contemporary literature, see the opening section of Salmon (1967).

9. Stroud (1981) well argues the case for a narrower, nonnaturalistic view of epistemology. A more naturalistic approach has long been advocated by Goldman (1978). Both of these essays are reprinted in (Kornblith 1985). A fuller treatment of Goldman's views may be found in his recent book (1986). To some extent my project for the philosophy of science is a special case of Goldman's project for epistemology as a whole.

10. The traditional philosopher's attitude toward the actual practice of science is illustrated by the following remark of Nelson Goodman's (1972, 168), quoted approvingly by Siegel (1980):

> The scientist may use platonistic class constructions, complex numbers, divination by inspection of entrails, or any claptrappery that he thinks may help him get the results he wants. But what he produces then becomes raw material for the philosopher, whose task is to make sense of all this: to clarify, simplify, explain, interpret in understandable terms. The practical scientist does the business but the philosopher keeps the books.

Goodman is extreme only in his arrogance.

11. For a more extensive discussion of the "biologizing" of Kant, see Lorentz (1941, 1962) and the essays in Wuketits (1984).

12. See Campbell (1960, 1974), Shimony (1971, 1981), Toulmin (1972), Hull (1978, 1982, forthcoming), and some of the essays in (Plotkin 1982). Bradie (1986) provides an informed, recent overview of evolutionary epistemologies.

13. I learned this argument from David Bloor in private conversation.

14. One can imagine a hypothetical Darwin setting out to create a theory of evolution by gathering together all such claims in the known literature and then attempting to test them against the empirical record. Such a Darwin would be as un-

known today as is Bergmann. One hundred such "rules" are to be found in Rensch (1971, 131–39). This reference I owe to David Hull, who generously loaned me his copy.

15. Some years ago I reviewed a volume from a conference on relationships between the history and the philosophy of science (Giere 1973a). At that time I concluded that the participants had failed to come to grips with the problem of how *descriptive* history could contribute to *normative* philosophy. The solution, I now think, is to abandon the philosophical search for normative principles of a categorical nature. That clears the way for a fruitful marriage between the two fields.

2. Theories of Science

1. Scholars are only now beginning to write in a serious way about the history of logical empiricism. Alberto Coffa's well informed, posthumous study of its intellectual origins and history through 1936 is in preparation. Oswald Haufling has published a useful book (1981a) and a collection of original papers (1981b) emphasizing the logical empiricist theory of language. Wesley Salmon's introduction to his (1979a) book provides a nice overview of Reichenbach's views. The best introduction to Carnap is his own "Intellectual Autobiography" (1963a). Popper's "Intellectual Autobiography" (1974) is less useful because it is excessively self-serving. Ayer (1959) contains a good selection of original papers. Finally, there is the two-volume collection edited by Neurath, Carnap, and Morris (1955, 1970), which contains all the articles of the original *International Encyclopedia of Unified Science*, including Joergen Joergensen's history of the movement up to World War II. It is a minor irony that Kuhn's *The Structure of Scientific Revolutions* first appeared as part of Volume 2 of that series. For post–World War II presentations and developments see Nagel (1961) and Hempel (1965).

2. Reichenbach clearly noted these three sources of logical empiricism in his one-sentence characterization of "logistic empiricism" in the preface to his 1938 book: "It is the intention of uniting both the empiricist conception of modern science and the formalistic conception of logic . . . which marks the working program of this philosophical movement."

3. Russell (1919), written while the author was in jail for his pacifism during World War I, contains a masterly, and largely nontechnical, presentation of this program.

4. Section 2 of Carnap (1963) contains a moving account of the inspiration Carnap received from reading Russell.

5. Here it is well to recall that *The Logic of Scientific Discovery* (1959a) is Popper's own translation of his 1934 work *Logic der Forschung*. Reichenbach's presentation of the distinction between discovery and justification appears in the first section of Reichenbach (1938). For a contemporary view of the distinction see the introduction and essays in Nickles (1980).

6. The authoritative source for the two-language view of theories is Carnap (1956). For an excellent survey of attempts to criticize the logical empiricist account of theories see the introduction to Suppe (1974).

7. For a review of more recent work on inductive logic see Giere (1979a).

8. This basic idea runs through Popper's writings from 1934 to the present. For a recent presentation see the essay "Conjectural Knowledge: My Solution to the Problem of Induction," reprinted in (Popper 1972, 1–31).

9. The authoritative source for Reichenbach's views is Reichenbach (1949), but the basic ideas are expressed much more simply in Salmon (1967).

10. The other philosopher who made much of the "long run" justification of science was the American pragmatist, C. S. Peirce.

11. This and others of Hempel's essays are reprinted in Hempel (1965). In addition to containing his own theory of explanation Salmon's recent book (1984) provides an authoritative review of philosophical work on explanation during the past twenty years. Achinstein (1983) provides a somewhat different perspective.

12. For a brief overview of Merton's work see Norman Storer's introduction to Merton (1973). Merton's papers on general sociological themes are reprinted in Merton (1949).

13. Originally published in 1942 as "Science and Technology in a Democratic Order," this essay was republished in Merton (1949) under the title "Science and Democratic Social Structure" and then reprinted in Merton (1973) under the title "The Normative Structure of Science." One cannot help wondering how much Merton's original analysis of the norms of science was influenced by a desire to show a correspondence between the ideals of science and the ideals of liberal democracy. The spector of Nazi Germany certainly looms large in this early essay.

14. This point of view is developed for the sociology of science in Mulkay (1979, 1980).

15. For an extended development of Mertonian ideas on the reward structure of science see Gaston (1978).

16. Major criticisms of various logical empiricist doctrines published around 1960 can be found in Hanson (1958) and Toulmin (1961). See also papers by Feyerabend (1962), Putnam (1962), Scriven (1958, 1962), and Sellars (1961). The collection of papers in Feigl and Maxwell (1961) gives a good picture of the state of the philosophy of science just before the appearance of Kuhn's book.

17. In his "Second Thoughts" (1974) Kuhn used the term "disciplinary matrix" in place of the more global sense of "paradigm."

18. For standard philosophical reactions to Kuhn's book see Shapere's review (1964) and Scheffler (1967).

19. Doppelt (1978) forcefully presents the idea that the important kind of incommensurability in Kuhn's account of science is not that of meaning but that of *standards*.

20. In "Second Thoughts" (1974) Kuhn himself can be seen to be moving in this direction.

21. Knowledgeable readers will be aware that other important proponents of alternative theories of science are being omitted. These include Feyerabend (1975, 1981), Shapere (1984), and Toulmin (1972). They are omitted not because I think these authors have nothing to contribute, but simply because their views, for what-

ever reasons, are not currently at the center of attention among people struggling with the problem of what a theory of science should be like.

22. One of my graduate students, Kevin Murnane, pointed out that the *initial* pursuit of any program must, by Laudan's criteria, be irrational because initially a program will have few if any solved problems and thus a very low, perhaps zero, rate of problem solving effectiveness.

23. Here Laudan refers to Popper's notorious theory of "verisimilitude," which has been reviewed, for example, by Newton-Smith (1981, chap. 8).

24. In addition to *The Scientific Image,* the reader might consult some of van Fraassen's papers such as those published in (1970, 1972, 1981). Van Fraassen (1985) is a lengthy reply to ten recent critics, including me.

25. Among anthologies containing works by recent sociologists of science are Barnes and Edge (1982), Barnes and Shapin (1979), Collins (1982), and Knorr-Cetina and Mulkay (1983b). For more critical collections, focusing on theoretical issues rather than empirical studies, see Hollis and Lukes (1982) and Brown (1984).

26. For a critique of interest theories from the standpoint of those more concerned with scientific discourse, see Woolgar (1981) and Yearley (1982).

3. Models and Theories

1. Here I am thinking of Stegmueller's (1979) "Bourbaki program" for reconstructing the physical sciences in a set-theoretical framework. The differences between Stegmueller's program and that of logical empiricism, as well as connections with the work of Patrick Suppes (1969), have been explored by Moulines and Sneed (1979).

2. The following remarks are based on an examination of roughly a dozen typical intermediate to advanced textbooks of classical mechanics. These include Abraham and Marsden (1978), Barger and Olsson (1973), Marion (1970), Sposito (1976), Synge and Griffith (1959), and Wallace and Fenster (1969), as well as several— Goldstein (1959), Slater and Frank (1947), Symon (1953)—I personally worked through either as an undergraduate or as a graduate student in physics. Undoubtedly, interesting things could be learned from a more systematic survey of texts in classical mechanics. But that is not necessary for the present inquiry. The features of classical mechanics that will be relevant here are obviously now so standardized in the textbook tradition that a broader sampling would not make much difference.

3. Indeed, my analysis in this chapter is very much in the spirit of Kuhn's writings, both the early and the late. My comments on the importance of specific force functions in mechanics, for example, parallel those found in Kuhn (1974).

4. Among the philosophers who have taken seriously issues concerning the use of idealizations and approximations in science are Scriven (1961), Shapere (1974, 1984), Suppe (1972), Moulines (1976), Laymon (1980), and Cartwright (1983).

5. When I say that hypotheses are linguistic entities, I mean this in the sense of what philosophers often call propositions as opposed to sentences. Sentences are written in a specific language using words in a particular order. Propositions are

more abstract. Different sentences, perhaps even in different languages, can express the same proposition.

6. The exception, as most philosophers will note, is the "redundancy theory of truth" according to which the content of any statement, *S,* is exactly the same as that of the metalinguistic statement "*S* is true." In this view the main function of the predicate "is true" is to facilitate "semantic ascent," that is, to talk about statements themselves rather than more directly about nonlinguistic objects.

7. It would be possible to interpret figure 3.10 along the lines of the theory of theories found in Suppes (1967, 1969), Sneed (1971), and Stegmueller (1976, 1979). The ovals could represent set-theoretical *predicates,* such as "is a classical particle system," rather than models. Despite its model-theoretic character, then, the Suppes, Sneed, Stegmueller view retains the linguistic flavor of the logical empiricist account. It is also part of a program of "rational reconstruction" that seems to me to be mainly an exercise in set theory.

8. The above paragraphs grew out of a conversation with David Hull, who insisted, from an evolutionary point of view, that it is descent, not similarity, that defines a biological species. Many months after this was written, I rediscovered a printed debate between Kuhn and Dudley Shapere over the same point (Suppe 1974, 506–9). Here, as in many matters, I side with Kuhn. Shapere (1985) seems still of the opinion that one can analyze science without any necessary reference to the judgments of individual scientists, a viewpoint I have explicitly criticized elsewhere (Giere 1985a).

9. Among the classic modern axiomatizations are Simon (1947, 1954) and McKinsey, Sugar, and Suppes (1953). Montague (1974) provided more than a hint of how horrendously complex a genuine first-order axiomatization might be.

10. For discussions of evolutionary theory substantially in agreement with my picture of theories, see Beatty (1981), Thompson (1983), and Lloyd (1984).

4. Constructive Realism

1. For a good introduction to the latest philosophical thinking on realism see the essays in Leplin (1984) or Churchland and Hooker (1985). For extended developments of individual viewpoints see Blackburn (1984), Devitt (1984), and Hooker (1987).

2. As a philosophical exercise one might consider whether the arguments van Fraassen uses against realism, and in favor of empiricism, might not also be used against his liberal empiricism (3) in favor of the more restricted versions (1) or (2). He would surely be uncomfortable with the more restrictive versions, but he perhaps cannot avoid them. This sort of argument has not been pushed against van Fraassen, perhaps simply because everyone is busy defending realism and no one wishes to defend more restrictive forms of empiricism.

3. The example of solar neutrinos, and its implications for philosophical discussions of observation, is nicely exploited by Shapere (1982).

4. Propensity interpretations of probability go back to Peirce, but their modern popularity began with Popper (1959b). I myself have developed propensity ideas in

a number of papers (1973b, 1976a, 1976b, 1979b, 1980). For a review from another perspective see Salmon (1979b).

5. In standard probability models the probability function takes real numbers as values. Ratios in finite populations are restricted to rational numbers. It is because of this discrepancy that most empiricist philosophers postulate limiting relative frequencies in infinite sequences to be the real world counterparts of probability models. The trouble, of course, is that the real world contains no infinite sequences. And so one invents hypothetical infinite sequences. Salmon (1967) provided a readable account of frequency interpretations of probability. Van Fraassen (1980, chap. 6) developed a more adequate, but less accessible, empiricist account.

6. Roger Shank, a specialist in artificial intelligence, has recently been giving lectures on "explanation." This is a natural extension of his earlier well-known work (Shank and Abelson 1977).

7. Salmon (1984, 119) considered the possibility that science might be primarily concerned with causal structure and not with individual events, and he therefore appealed to the existence of "applied science" to justify a concern with explaining individual events. But this put him in the awkward position of seeking to "understand scientific understanding" largely through the study of explanations in applied science. My hypothesis as to why Salmon has not been more bothered by this discrepancy is that his real concern is less with *understanding* scientific explanations than with *justifying* them. His main concern throughout is with the characteristics that distinguish a legitimate scientific explanation from nonscientific, or pseudoscientific, explanations. Thus, for Salmon, as for logical empiricists generally, the study of explanation is another means toward the overall goal of providing a philosophical justification for science. Recall that Salmon's earlier work (1967) focused on the traditional problem of justifying induction.

5. Realism in the Laboratory

1. One might well ask why I choose this laboratory for study. I did so for several reasons. Shortly after I became interested in the possibility of laboratory studies, I was invited by an academic acquaintance to serve as a "humanistic" consultant for a film on research at IUCF. That project provided my introduction to the scientists at the facility. As it happens, IUCF is the largest laboratory associated with Indiana University. In addition, I already had some familiarity with nuclear physics, gained as a graduate student in physics two decades earlier. These, then, were the major "contingencies" in my own research.

I began observing operations at IUCF in the fall of 1983 and continued to do so until the end of 1986. I received much help from two "native informants"— physicists who work full-time at the facility. They allowed me to follow them around and patiently answered my questions. Occasionally, I tape-recorded our conversations. I saw experiments being set up, equipment being built, and data being gathered in the middle of the night. I heard presentations of preliminary findings and attended PAC meetings where proposals were presented and evaluated. I have no special training in ethnographic methods, but I have not attempted an ethnography of the laboratory. My aims were much more limited. For the achievement of those aims my understanding of the basic physics involved seemed much more valuable than any ethnographic techniques I might have learned in preparation for my observations.

2. There is a great deal of similarity between the argument of this section and that in the last chapter of Ian Hacking's recent book (1983). In fact, I originally developed these ideas before I saw Hacking's book. I first presented them publicly in Ghent, Belgium, at a meeting of the Society for Social Studies of Science in November 1984. An abstract, titled "Realism in the Laboratory," which was written six months earlier, appeared in the proceedings (Giere 1984b). Hacking obviously came to similar conclusions much earlier, and I have now benefited from his account. What we primarily share is the inspiration of having spent time in a modern scientific laboratory. I do not know what prompted Hacking to do so. I was prodded by reading Latour and Woolgar (1979) and Knorr-Cetina (1981).

3. I think this conclusion is in the spirit of Arthur Fine's "Natural Ontological Attitude" (1984a, 1984b). Perhaps a naturalistic constructive realism can be seen as an extension of the natural ontological attitude to the activity of science itself. Fine has pointed out to me that something like constructive realism, complete with the phrase "family of models," appears in one of his earlier essays (1975, 28–30).

4. Here I am not merely surmising what van Fraassen would say but reporting on what he did say when I presented this material in March 1986 at a workshop sponsored by the Center for Philosophy of Science at the University of Minnesota. I thank Philip Kitcher, Wade Savage, and, of course, Bas van Fraassen, for having provided the occasion for that exchange.

5. This quotation is a transcription of a discussion between Karen Knorr-Cetina and myself recorded for Belgian radio in November 1984. I thank Werner Callebaut both for having made the original recording and for having provided me with a copy.

6. In addition to having been described in Evans-Pritchard (1937), the Azande chicken oracle was discussed, for example, in Barnes (1974), Bloor (1976), and Collins and Pinch (1982). It has also been the subject of much debate among philosophers and anthropologists concerned with the "rationality" of primitive peoples. For a selection of relevant papers see Wilson (1970) and Hollis and Lukes (1982).

7. I cannot resist recording the suggestion that one field that arguably exemplifies a constructivist model is constructivist sociology of science itself. This suggestion should please at least those constructivists who value irony and self-reference (for example, Woolgar 1983).

8. Perhaps one of the reasons progress in the social sciences has been so slow is that we humans have evolved no mechanisms for dealing with large social organizations. What evolution provided us was a capacity for dealing face-to-face with individuals in small groups.

9. The most consistent advocate of the importance of instrumentation in science was the late Derek Price (1965, 1984). In addition to Hacking (1983) others who

have recently taken up this theme include Ackermann (1985), Franklin (1986), and Galison (1985, 1987a, 1987b).

6. Scientific Judgment

1. Standard references include Nisbett and Ross (1980) and Kahneman, Slovic, and Tversky (1982). For recent reviews of the literature see Einhorn and Hogarth (1981) and Pitz and Sachs (1984).

2. The classic treatment is von Neumann and Morgenstern (1944). For a standard text see Luce and Raiffa (1957).

3. For those who prefer symbols to words, the following is a version of the relevant axioms and theorems (Winkler 1972).

AXIOM 1. *(Comparability) For any two outcomes, O_i and O_j, either O_i is preferred to O_j or O_j is preferred to O_i, or neither is preferred to the other.*

AXIOM 2. *(Transitivity) If O_i is preferred to O_j, and O_j is preferred to O_k, then O_i is preferred to O_k.*

AXIOM 3. *(Continuity) If O_i is preferred to O_j, and O_j is preferred to O_k, then there are two p-mixtures (different p's) of O_i and O_k such that O_j is preferred to the first but not to the second.*

AXIOM 4. *(Admissibility) If O_i is preferred to O_j, then for any other outcome, O_k, the p-mixture of O_i and O_k is preferred to the p-mixture (same p) of O_j and O_k.*

AXIOM 5. *(Substitutability) If the agent is indifferent to the choice between outcomes O_j and O_k, then they may be interchanged in any decision problem.*

AXIOM 6. *(Monotonicity) If O_i is preferred to O_j, then the p-mixture of O_i and O_j is preferred to their q-mixture if and only if p is greater than q.*

THEOREM 1. *The agent's preferences for the outcomes are characterized by a utility function, $U(O_i)$, such that:*
(a) $U(O_i) > U(O_j)$ if and only if O_i is preferred to O_j.
(b) If the agent is indifferent to the choice between O_i and a p-mixture of O_j and O_k, then $U(O_i) = p\, U(O_j) + (1 - p)\, U(O_k)$.

THEOREM 2. *There is a function P defined over the states of the world, S, T, U, etc., such that:*
(a) 0 is less than or equal to P(S) is less than or equal to 1,
(b) $P(S \text{ or } T) = P(S) + P(T)$,
(c) $P(S \text{ or Not-}S) = 1$.

THEOREM 3. *The agent prefers option A to option B if and only if EU(A) is greater than EU(B).*

4. It is somewhat surprising that satisficing models are not more studied by economists or psychologists (or philosophers!). An exception is Wimsatt (1980). Part of the reason for their neglect may be that satisficing models do not lend themselves to a self-contained mathematical treatment like that which defines Bayesian agents. The mathematical characterization of satisficing is relatively trivial. All that is of the interest is in the applications.

5. The term *open-minded* was first used in a similar context by Abner Shimony

(1970), who made it part of his "tempered personalism" that no hypothesis be assigned a prior probability of one or zero.

6. This argument is implicit in *The Scientific Image* (1980) and much more explicit in a later paper (van Fraassen 1981). I have previously discussed this argument at somewhat greater length (Giere 1985b).

7. Boyd (1973, 1981) has long emphasized the role of background information in a realistic interpretation of scientific inference.

8. It is interesting to recall that long ago Carnap (1963b) thought of inductive logic in terms of the design of an ideal robot.

7. Models and Experiments

1. Concurrently with my study of (p,n) experiments discussed in chapter 5, I also followed the series of elastic scattering experiments reported in the present chapter. Overall, I followed these developments for about two years. In addition, I read in the relevant experimental and theoretical literature. My own research was facilitated by several members of the experimental staff at the Indiana University Cyclotron Facility, one of whom I formally interviewed (with tape recorder running) on three different occasions. I also interviewed two other experimentalists and one theoretician at IUCF. In the summer of 1986 I attended a conference devoted entirely to theoretical aspects of the subject, although many of the key experimentalists were also present. On that occasion I was able to interview formally several key people I had not previously met. In quoting from my interviews, I have identified the most prominent people by name because anyone who knows the field could not fail to identify them. The others I have not identified.

2. For a more extended discussion of the role of predictions in testing theories, including historical references, see Giere (1983). The strategy of that treatment, however, is different from the present one. There I sought to justify a belief in the power of successful predictions using satisficing as a decision strategy. Here I use the fact that scientists honor the power of predictions as evidence that they are satisficers. The difference is whether one seeks to justify science or to explain it.

3. It is particularly appropriate that a satisficing strategy should explain why prediction matters in judgments about models because Simon, who has championed satisficing, was once also concerned with the confirmatory role of predictions (Simon 1955), although he did not connect it with satisficing. I distinctly remember reading Simon's 1955 paper as a graduate student over 20 years ago.

4. There were at least two groups working on the relativistic impulse approximation at that time. Another consisted of Clark herself and her longtime collaborators; Lanny Ray, a member of the original Texas group; and Brian Serot, one of Dirk Walecka's students. Although there was much communication among all the parties, at some point a few of the participants began see the situation as a race to develop the relativistic impulse approximation. This case thus nicely illustrates the delicate balance between cooperation and competition in science. But that is not my focus here. All the quotations that follow are from a joint interview with Wallace, McNeil, and Shepard conducted in the summer of 1986.

5. Shepard's remark that the fit got better as the idealizations were relaxed illustrates a point long emphasized by Ron Laymon (1980, 1982).

6. One can find references to "cognitive resources" in the recent sociological literature (Mulkay 1979, 93), but there they are assumed to be the common property of a scientific subculture, a resource to which all members can appeal. Pickering (1984) talked about the "resource of expertise" as part of his idea of "opportunism in context." That is very close to my idea of an individual's "cognitive resources." DeMey used the term "cognitive resource" for individuals (1982, 201) but has not developed the idea. There is also some overlap between the idea of an individual's cognitive resources and Howard Gruber's (1980) notion of an individual's "network of enterprises."

7. Those familiar with the sociological literature on marginal innovation in science will immediately recognize Clark as a paradigm of the "marginal" person. (Gieryn and Hirsh [1983] provide a quick overview of this literature.) She is a woman in a field that is 95 percent male; she approaches models "phenomenologically" rather than from more fundamental principles; and despite having begun her research in earnest in 1961, only acquired a full-time position as an assistant professor in 1981. Indeed, she herself cited her marginal status (not by that name, of course) as one of the reasons she succeeded: "I was fortunate, I guess, that I didn't have the kind of job that said, 'If you don't publish what people tell you you should be working on, you know, your going to get fired'."

8. There are similarities here with Pickering's (1981, 1984) analysis of the 1970s revolution in high-energy physics.

8. Explaining the Revolution in Geology

1. I have consulted some primary sources, talked to a number of geologists, and even formally interviewed several myself. Nevertheless, this chapter relies heavily on secondary literature, including works by historians and philosophers of science who have recently turned their attention to this subject (R. Laudan 1981; Ruse 1981). I am particularly indebted to Henry Frankel (1980b, 1981, 1982, 1987), who provided my first serious introduction to the subject. He also generously provided a number of useful comments on this chapter. I have also borrowed much from Glen's (1982) work. Frankel (1983) claims that Glen's overall presentation is skewed too much toward the contributions of Cox's group in California. On this point I agree with Frankel. My treatment makes much use of Glen's *data*, particularly his extensive interviews, which make a cognitive reconstruction possible. My great debt to Glen will be obvious. The *interpretation* of Glen's data, however, is my own.

2. My first serious introduction to the 1926 symposium was through reading the master's dissertation of Raymond Schlosser (1984). The symposium has been discussed in several published sources, including Marvin (1973) and Hallam (1973).

3. Rollin Chamberlin provides a striking example of how little effect the methodological pronouncements of scientists may have on the actual practice of science. In 1890 the elder Chamberlin, Rollin's father, published a paper entitled "The Method of Multiple Working Hypotheses," in which he argued the methodological virtues

of pursuing several different hypotheses simultaneously (T. C. Chamberlin 1890). And, as Rachel Laudan (1980) has argued, he seems to have meant to include widely divergent hypotheses, as widely divergent as mobilism and stabilism. Thirty-six years after the appearance of this paper, there was no hint of such an attitude in Rollin Chamberlin's reaction to Wegener's hypothesis. In fact, just the reverse. He was far more concerned to defend his father's theory than his father's methodological pronouncements. My suspicion is that the virtues of multiple working hypotheses tend to be advanced by scientists attempting to gain a hearing for a new view, as was apparently the case with the elder Chamberlin. Correspondingly, those virtues tend to be downplayed by scientists, like the younger Chamberlin, defending an established position.

4. Stewart's article (1986) is but a tiny sample of his forthcoming book. I thank him for the opportunity to see the whole manuscript before publication.

5. My knowledge of directional geomagnetism owes much to the work of my student, Jeffrey Whitmer, whose dissertation in progress examines the implications of this episode for current philosophical theories of scientific development.

6. The importance of technology for the geosciences is well recognized by the geoscience community itself. For example, in 1979 the Geophysics Study Committee of the National Research Council sponsored a study published under the title *The Impact of Technology on Geophysics* (Newell et al. 1979).

7. Hess's paper was circulated in 1960 but delayed in publication until 1962. In the meantime Dietz (1961) published the idea and introduced the expression "seafloor spreading." Dietz, however, was a bit tardy in acknowledging his debt to Hess (Dietz 1968). The reason Vine and Matthews referred to Dietz rather than Hess was simply that the Hess paper had not yet appeared.

8. Lawrence Morley submitted a clear statement of the Vine-Matthews hypothesis in the form of a letter to *Nature* in February 1963, several months before submission of the Vine-Matthews paper. It was rejected. Morley is reported as claiming that the reader for *Nature* had heard that the Vine-Matthews paper was soon to be submitted and rejected his letter so that Vine and Matthews would get priority (Frankel 1982, 16–17). There are other possible explanations. Frankel and Glen insist on referring to the hypothesis as the Vine-Matthews-Morley hypothesis. I have not followed this suggestion because it does not square with my more biological (and less intellectual) theory of science. In a biological theory it is to be expected that different researchers should independently hit on structurally similar models. And examples of that phenomenon are worth noting in the study of science. What matters to the science itself, however, is the causal influence of a model in the subsequent evolution of the field. Morley's construction of the model, as everyone admits, had absolutely no effect on subsequent developments. That was partly because Morley was at the time an administrator with the Geological Survey of Canada and had little time to pursue, or to push, his ideas. Vine did. In science, as in biology, it is *success,* and not just *potential,* that counts.

9. There is some controversy among commentators such as Glen (1982) and Frankel (1982) over when Vine actually accepted seafloor spreading as opposed to merely pursuing it. My view is that such questions often have no definitive answer.

The psychological reality may be less clear-cut than this distinction implies. The low point in Vine's confidence was in late 1964 before Hess and Wilson appeared on the scene in Cambridge. Their joint analysis of the Juan de Fuca Ridge surely increased his confidence. Vine's reported comments strongly suggest that he came to full conviction upon learning of the existence of the Jaramillo event. That evened out the time scale so that the spreading rate is nearly constant. In any case, there is no question that by the time he had seen the *Eltanin*-19 profile and Opdyke's core analysis, the decision had been made.

10. It will be noted that my characterization of a "scientific revolution" here is primarily social. A revolution is completed when the vast majority of scientists in the relevant communities adopt the new view. This seems to me the only viable characterization. What sense would it make to invoke some other criterion if in fact most of the relevant scientists do not agree with the conclusion?

11. From my earlier writings (Giere 1983, 1984a, 1985b) as well as from the previous chapters, one may justifiably think that I am here putting special significance on the fact that the character of magnetic profiles like those brought back by the *Eltanin* had been *predicted* by Vine and Matthews. One may even assume I think it significant that this was a prediction in the literal sense of being pronounced before having found the evidence. As a matter of fact, I think there *are* cases in which the existence of a prediction in the literal sense is relevant to the judgment of how likely the observed evidence would be if the proposed hypothesis were mistaken. In this particular case, however, the fact that there was a literal prediction seems not to have been important. What mattered most was the detail in the data together with the model-based judgment that such data were fantastically improbable in any stabilist model. Indeed, one can imagine a scenario in which Vine missed Hess's lecture and was assigned to someone other than Matthews as a graduate student, so that there was no Vine-Matthews hypothesis. Nevertheless, Pitman produces the *Eltanin*-19 profile, Opdyke analyzes the data from deep-sea cores, and someone else, perhaps Opdyke himself, comes up with the model that in fact was proposed by Vine and Matthews. In this scenario the "revolution" might have been delayed by six months to a year, but things now, 20 years later, would be little changed.

12. The closest Glen came to providing a theoretical explanation of these developments in the earth sciences was in a footnote (Glen 1982, 314) in which he appealed to Hudson's (1966) distinction between "convergers" and "divergers." He also referred to works by McClelland (1970), and Mitroff (1974). All those works, which analyzed the psychology of scientists in terms of personality types, antedated the recognition of the cognitive sciences in the contemporary sense. Now one would talk about cognitive resources, or perhaps cognitive style, and judgment. This preserves what relevance there is in personality types, while focusing on the kinds of cognitive mechanisms that make possible good scientific explanations of how science evolves.

13. In a forthcoming article Rachel and Larry Laudan argue that differences in judgment among earth scientists at various times between the mid-1950s and the mid-1960s are to be explained by differences in methodological standards. General agreement was reached not because of agreement on standards but because, with

the new evidence of the mid-1960s, mobilism had come to satisfy everyone's standards for an acceptable theory. My view, of course, is that methodological standards provide little explanation of scientists' decisions—even when cited as such by scientists themselves. One wants to explain the standards in terms of more basic decision strategies. My earlier explanation of the appeal to successful "prediction" is an example of the kind of explanation I have in mind. It also has the advantage that it allows a proper role for nonepistemic interests.

Epilogue: Reflexive Reflections

1. Given the intellectual background, the choice of title for this epilogue is embarrassingly obvious. After I had chosen it, I discovered it has also been used by Jay Stewart in his as yet unpublished manuscript. He no doubt thought of it first, but I cannot resist employing myself.

References

Abraham, R., and J. E. Marsden. 1978. *Foundations of mechanics.* 2d ed. Reading, Mass.: Benjamin.

Achinstein, P. 1983. *The nature of explanation.* New York: Oxford University Press.

Ackermann, R. J. 1985. *Data, instruments, and theory.* Princeton: Princeton University Press.

Adie, R. J. 1952. The position of the Falkland Islands in a reconstruction of Gondwanaland. *Geological Magazine* 89:401–10.

Arbib, M. A., and M. B. Hesse. 1986. *The construction of reality.* Cambridge: Cambridge University Press.

Ayer, A. J. 1959. *Logical positivism.* New York: Free Press.

Barger, V., and M. Olsson. 1973. *Classical mechanics: A modern perspective.* New York: McGraw-Hill.

Barnes, B. 1974. *Scientific knowledge and sociological theory.* London: Routledge & Kegan Paul.

———. 1982. *T. S. Kuhn and social science.* New York: Columbia University Press.

Barnes, B., and D. Bloor. 1982. Relativitism, rationalism and the sociology of knowledge. In *Rationality and relativism,* ed. M. Hollis and S. Lukes. Cambridge: MIT Press.

Barnes, B., and D. Edge, eds. 1982. *Science in context.* Cambridge: MIT Press.

Barnes B., and S. Shapin, eds. 1979. *Natural order: Historical studies of scientific culture.* Beverly Hills, Calif.: Sage.

Bartlett, F. C. 1932. *Remembering.* Cambridge: Cambridge University Press.

Beatty, J. 1981. What's wrong with the received view of evolutionary theory? In *PSA 1980,* vol. 2, ed. P. D. Asquith and R. N. Giere, 397–426. East Lansing, Mich.: Philosophy of Science Association.

Berger, P., and T. Luckmann. 1966. *The social construction of reality.* New York: Doubleday.

Berlin, B., and P. Kay. 1969. *Basic color terms: Their universality and evolution.* Berkeley: University of California Press.

Blackburn, S. 1984. *Spreading the word.* Oxford: Clarendon Press.

Blackett, P. M. S., E. C. Bullard, and S. K. Runcorn, eds. 1965. *A symposium on continental drift.* Philosophical Transactions of the Royal Society of London, ser. A, vol. 258.

Bloor, D. 1976. *Knowledge and social imagery.* London: Routledge & Kegan Paul.

———. 1981. The strengths of the strong programme. *Philosophy of the Social Sciences* 11:199–213.

———. 1982. Durkheim and Mauss revisited: Classification and the sociology of knowledge. *Studies in the History and Philosophy of Science* 13:267–97.

Boyd, R. 1973. Realism, underdetermination, and a causal theory of reference. *Nous* 7:1–12.

———. 1981. Scientific realism and naturalistic epistemology. In *PSA 1980,* vol. 2, ed. P. D. Asquith and R. N. Giere, 613–52. East Lansing, Mich.: Philosophy of Science Association.

———. 1984. The current status of scientific realism. In *Scientific Realism,* ed. J. Leplin, 41–82. Berkeley: University of California Press.

Bradie, M. 1986. Assessing evolutionary epistemologies. *Biology and Philosophy* 1:401–59.

Bromberger, S. 1963. A theory about the theory of theory and about the theory of theories. In *Philosophy of science: The Delaware Seminar,* vol. 2, ed. B. Baumrin. New York: Wiley.

Brown, J. R., ed. 1984. *Scientific rationality: The sociological turn.* Dordrecht: Reidel.

Bucher, W. H. 1933. *The deformation of the earth's crust.* Princeton: Princeton University Press.

Bullard, E. C., J. E. Everett, and A. G. Smith. 1965. Fit of continents around the Atlantic. In *A symposium on continental drift,* ed. P. M. S. Blackett, E. C. Bullard, and S. K. Runcorn, 41–75. Philosophical Transactions of the Royal Society of London, ser. A, vol. 258.

Campbell, D. T. 1959. Methodological suggestions from a comparative psychology of knowledge processes. *Inquiry* 2:152–82.

———. 1960. Blind variation and selective retention in creative thought as in other knowledge processes. *Psychological Review* 67:380–400.

———. 1966. Pattern matching as essential in distal knowing. In *The psychology of Egon Brunswick,* ed. K. R. Hammond. New York: Holt, Rinehart & Winston. Reprinted in Kornblith (1985).

———. 1974. Evolutionary epistemology. In *The philosophy of Karl Popper,* ed. P. A. Schilpp. La Salle, Ill.: Open Court.

———. 1985. Science policy from a naturalistic sociological epistemology. In *PSA 1986,* vol. 2, ed. P. D. Asquith and P. Kitcher, 14–29. East Lansing, Mich.: Philosophy of Science Association.

Carnap, R. 1928. *Der logische Aufbau der Welt.* Berlin: Welkreis.

———. 1937. *The logical syntax of language.* London: Routledge & Kegan Paul.

―――. 1950. *Logical foundations of probability.* Chicago: University of Chicago Press (2d ed., 1962).

―――. 1956. The methodological character of theoretical concepts. In *The foundations of science and the concepts of psychology and psychoanalysis,* ed. H. Feigl and M. Scriven. Minnesota Studies in the Philosophy of Science, vol. 1. Minneapolis: University of Minnesota Press.

―――. 1963a. Intellectual autobiography. In *The philosophy of Rudolf Carnap,* ed. P. A. Schilpp. La Salle, Ill.: Open Court.

―――. 1963b. The aim of inductive logic. In *Logic, methodology, and the philosophy of science,* vol. 1, ed. E. Nagel, P. Suppes, and A. Tarski, 303–18. Stanford: Stanford University Press.

―――. 1967. *The logical construction of the world,* trans. R. A. George. Berkeley: University of California Press.

Cartwright, N. 1983. *How the laws of physics lie.* Oxford: Clarendon Press.

Casscells, W., et al. 1978. Interpretation by physicians of clinical laboratory results. *New England Journal of Medicine* 299:999–1000.

Chamberlin, T. C. 1890. The method of multiple working hypotheses. *Science* 15:92–96.

Churchland, P. M. 1979. *Scientific realism and the plasticity of mind.* Cambridge: Cambridge University Press.

―――. 1986. Cognitive neurobiology: A computational hypothesis for laminar cortex. *Biology and Philosophy* 1:25–51.

Churchland, P. M., and C. A. Hooker, eds. 1985. *Images of science.* Chicago: University of Chicago Press.

Churchland, P. S. 1986. *Neurophilosophy.* Cambridge: MIT Press.

Clark, B. C., S. Hama, and R. L. Mercer. 1983. Dirac phenomenology and the Nuclear Optical Model. In *The interaction between medium energy nucleons in nuclei—1982,* ed. H. O. Meyer. New York: American Institute of Physics.

Collins, H. M. 1981a. Stages in the empirical program of relativism. *Social Studies of Science* 11:3–10.

―――. 1981b. The role of the core-set in modern science: Social contingency with methodological propriety in science. *History of Science* 19:6–19.

―――. ed. 1982. *Sociology of scientific Knowledge—A source book.* Bath: Bath University Press.

Collins, H. M., and G. Cox. 1976. Recovering relativity: Did prophecy fail? *Social Studies of Science* 6:423–44.

Collins, H. M., and T. J. Pinch. 1982. *Frames of meaning.* London: Routledge & Kegan Paul.

Cox, A. 1969. Geomagnetic reversals. *Science* 163:237–45. Reprinted in Cox (1973).

Cox, A., ed. 1973. *Plate tectonics and geomagnetic reversals.* San Francisco: Freeman.

Cox, A., R. R. Doell, and G. B. Dalrymple. 1963. Geomagnetic polarity epochs and pleistocene geochronometry. *Nature* 198:1049–51. Reprinted in Cox (1973).

————. 1964. Reversals of the earth's magnetic field. *Science* 144:1537–43. Reprinted in Cox (1973).

DeMey, M. 1982. *The cognitive paradigm*. Dordrecht: Reidel.

Devitt, M. 1984. *Realism and truth*. Princeton: Princeton University Press.

Dietz, R. S. 1961. Continent and ocean basin evolution by spreading of the sea floor. *Nature* 190:854–57.

————. 1968. Reply. *Journal of Geophysical Research* 73:6567.

Doell, R. R., and G. B. Dalrymple. 1966. Geomagnetic polarity epochs: A new polarity event and the age of the Brunhes-Matuyama boundary. *Science* 152:1060–61. Reprinted in Cox (1973).

Doppelt, G. 1978. Kuhn's epistemological relativism: An interpretation and defense. *Inquiry* 21:33–86.

————. 1986. Relativism and the reticulational model of scientific rationality. *Synthese* 69:225–52.

Dorling, J. 1972. Bayesianism and the rationality of scientific inference. *British Journal for the Philosophy of Science* 23:181–90.

Douglas, M. 1970. *Natural symbols*. London: Barrie & Jenkins.

Duhem, P. 1914. *La théorie physique*. Paris. English edition, *The aim and structure of physical theory,* trans. P. P. Wiener. Princeton: Princeton University Press, 1954.

Durkheim, E. 1915. *The elementary forms of the religious life,* trans. J. W. Swain. London: Allen & Unwin.

Du Toit, A. L. 1937. *Our wandering continents*. Edinburgh: Oliver and Boyd.

Edge, D., and M. Mulkay. 1976. *Astronomy transformed*. New York: Wiley.

Edwards, W. 1968. Conservatism in human information processing. In *Formal representation of human judgment,* ed. B. Kleinmuntz. New York: Wiley.

Einhorn, H. J., and R. M. Hogarth. 1981. Behavioral decision theory: Processes of judgment and choice. *Annual Review of Psychology* 32:53–88.

Ericsson, K. A., and H. A. Simon. 1984. *Protocol analysis*. Cambridge: MIT Press.

Evans-Pritchard, E. E. 1937. *Witchcraft, oracles, and magic among the Azande*. Oxford: Clarendon Press.

Farley, J., and G. Gieson. 1974. Science, politics, and spontaneous generation in nineteenth-century France: The Pasteur-Pouchet debate. *Bulletin of the History of Medicine* 48:161–98.

Faust, D. 1984. *The limits of scientific reasoning*. Minneapolis: University of Minnesota Press.

Feigl, H. 1970. The "orthodox" view of theories: Remarks in defense as well as critique. In *Analysis of theories and methods of physics and psychology,* ed. M. Radner and S. Winokur. Minnesota Studies in the Philosophy of Science, vol. 4. Minneapolis: University of Minnesota Press.

Feigl, H., and G. Maxwell, eds. 1961. *Current issues in the philosophy of science*. New York: Holt, Rinehart & Winston.

Feyerabend, P. K. 1962. Explanation, reduction, and empiricism. In *Scientific explanation, space, and time,* ed. H. Feigl and G. Maxwell. Minnesota Studies in the Philosophy of Science, vol. 3. Minneapolis: University of Minnesota Press.

————. 1975. *Against method*. London: New Left Books.

————. 1981. *Philosophical Papers*, 2 vols. Cambridge: Cambridge University Press.

Fine, A. 1975. How to compare theories: Reference and change. *Nous* 9:17–32.

————. 1984a. The natural ontological attitude. In *Scientific realism*, ed. J. Leplin. Berkeley: University of California Press. Reprinted in Fine (1986).

————. 1984b. And not anti-realism either. *Nous* 18:51–65. Reprinted in Fine (1986).

————. 1986. *The shaky game: Einstein, realism, and the quantum theory*. Chicago: University of Chicago Press.

Frankel, H. 1978. Arthur Holmes and continental drift. *British Journal for the History of Science* 11:130–50.

————. 1980a. Hess's development of his seafloor spreading hypothesis. In *Scientific discovery*, ed. T. Nickles, 345–66. Dordrecht: Reidel.

————. 1980b. Problem solving, research traditions, and the development of scientific fields. In *PSA 1980*, vol. 1, ed. P. D. Asquith and R. N. Giere, 29–40. East Lansing, Mich.: Philosophy of Science Association.

————. 1981. The non-Kuhnian nature of the recent revolution in the earth sciences. In *PSA 1978*, vol. 2, ed. P. D. Asquith and I. Hacking, 197–214. East Lansing, Mich.: Philosophy of Science Association.

————. 1982. The development, reception, and acceptance of the Vine-Matthews-Morley hypothesis. *Historical Studies in the Physical Sciences* 13:1–39.

————. 1983. Review of W. Glen, *The road to Jaramillo*. EOS 64:394–96.

————. 1987. The continental drift debate. In *Scientific controversies*, ed. H. T. Engelhardt, Jr., and A. L. Caplan, 203–48. Cambridge: Cambridge University Press.

Franklin, A. 1986. *The neglect of experiment*. Cambridge: Cambridge University Press.

Friedman, M. 1983. *Foundations of space-time theories*. Princeton: Princeton University Press.

Gale, G. 1984. Science and the philosophers. *Nature* 132:491–95.

Galison, P. 1985. Bubble chambers and the experimental workplace. In *Observation, experiment, and hypothesis in modern physical science*, ed. P. Achinstein and O. Hannaway, 309–73. Cambridge: MIT Press.

————. 1987a. Artificial clouds, real particles. In *The uses of experiment*, ed. D. Gooding. Cambridge: Cambridge University Press.

————. 1987b. Bubbles, sparks and the postwar laboratory. In *Physics in the 1950s*, ed. L. Hoddeson, L. Brown, and M. Dresden. Cambridge: Cambridge University Press.

Gallistel, C. R. 1980. *The organization of action: A new synthesis*. Hillsdale, N.J.: Erlbaum.

Gamow, G. 1947. *One, two, three . . . infinity*. New York: Viking Press.

Gardner, H. 1985. *The mind's new science*. New York: Basic Books.

Garfinkel, H. 1967. *Studies in ethnomethodology*. Englewood Cliffs, N.J.: Prentice-Hall.

Garland, G. D., ed. 1966. *Continental drift*. Toronto: University of Toronto Press.

Gaston, J. 1978. *The reward system in British and American science.* New York: Wiley.

Georgi, J. 1962. Memories of Alfred Wegener. In *Continental drift,* ed. S. K. Runcorn, 309–24. New York: Academic Press.

Giere, R. N. 1973a. History and philosophy of science: Intimate relationship or marriage of convenience? *British Journal for the Philosophy of Science* 24: 282–97.

————. 1973b. Objective single case probabilities and the foundations of statistics. In *Logic, methodology, and philosophy of science,* vol. 4, ed. P. Suppes et al., 467–83. Amsterdam: North-Holland.

————. 1976a. A Laplacean formal semantics for single-case propensities. *Journal of philosophical logic* 5:321–53.

————. 1976b. Empirical probability, objective statistical methods, and scientific inquiry. In *Foundations of probability theory, statistical inference, and statistical theories of science,* vol. 2, ed. C. A. Hooker and W. Harper, 63–101. Dordrecht: Reidel.

————. 1977. Testing vs. information models of statistical inference. In *Logic, laws, and life,* ed. R. G. Colodny, 19–70. Pittsburgh Series in the Philosophy of Science, vol. 6. Pittsburgh: University of Pittsburgh Press.

————. 1979a. Foundations of probability and statistical inference. In *Current research in philosophy of science,* ed. P. D. Asquith and H. Kyburg, Jr., 503–33. East Lansing, Mich.: Philosophy of Science Association.

————. 1979b. Propensity and necessity. *Synthese* 40:439–51.

————. 1980. Causal systems and statistical hypotheses. In *Applications of inductive logic,* ed. L. J. Cohen and M. B. Hesse, 251–70. Oxford: Oxford University Press.

————. 1983. Testing theoretical hypotheses. In *Testing scientific theories,* ed. J. Earman, 269–98. Minnesota Studies in the Philosophy of Science, vol. 10. Minneapolis: University of Minnesota Press.

————. 1984a. Toward a unified theory of science. In *Science and reality,* ed. J. T. Cushing, C. F. Delaney, and G. M. Gutting, 5–31. Notre Dame, Ind.: University of Notre Dame Press.

————. 1984b. Realism in the laboratory. In *George Sarton centennial,* ed. W. Callebaut et al. Ghent, Belgium: Communication and Cognition, 15–18.

————. 1985a. Background knowledge in science: A naturalistic critique. In *PSA 1984,* vol. 2, ed. P. Asquith and P. Kitcher, 664–71. East Lansing, Mich.: Philosophy of Science Association.

————. 1985b. Constructive realism. In *Images of science,* ed. P. M. Churchland and C. A. Hooker. Chicago: University of Chicago Press.

Gieryn, T. F., and R. F. Hirsh. 1983. Marginality and innovation in science. *Social Studies of Science* 13:87–106.

Gilbert, G. N. 1980. Being interviewed: A role analysis. *Social Science Information* 19:227–36.

Gilbert, G. N., and M. Mulkay. 1981. Contexts of scientific discourse: Social accounting in experimental papers. In *The social process of scientific investiga-*

tion, ed. K. Knorr-Cetina et al., 269–94. Sociology of the Sciences Yearbook, vol. 4. Dordrecht: Reidel.

———. 1982. Warranting scientific belief. *Social Studies of Science* 12:383–408.

———. 1984. *Opening Pandora's box.* Cambridge: Cambridge University Press.

Glen, W. 1982. *The road to Jaramillo.* Stanford: Stanford University Press.

Goldman, A. I. 1978. Epistemics: The regulative theory of cognition. *The Journal of Philosophy* 75:509–23. Reprinted in Kornblith (1985).

———. 1986. *Epistemology and cognition.* Cambridge: Harvard University Press.

Goldstein, H. 1959. *Classical mechanics.* New York: Addison-Wesley.

Goodman, N. 1970. Seven strictures on similarity. In *Experience and theory,* ed. L. Foster and J. W. Swanson. Boston: University of Massachusetts Press. Reprinted in Goodman (1972).

———. 1972. *Problems and projects.* Indianapolis: Bobbs-Merrill.

Green, A. E. S. 1985. Relativistic scalar-vector potentials for *N-N* and *N-A* interactions, Rome revisited. In *Antinucleon- and nucleon-nucleus interactions,* ed. G. E. Walker, C. D. Goodman, and C. Olmer, 143–57. New York: Plenum.

Griggs, D. T. 1939. A theory of mountain building. *American Journal of Science* 237:611–50.

Griggs, R. A., and S. E. Ransdell. 1986. Scientists and the selection task. *Social Studies of Science* 16:319–30.

Gruber, H. E. 1980. Cognitive psychology, scientific creativity, and the case study method. In *On scientific discovery,* ed. M. D. Grmek, R. S. Cohen, and G. Cimino, 295–322. Dordrecht: Reidel.

———. 1981. *Darwin on man.* Chicago: University of Chicago Press.

Gruber, H. E., and J. J. Voneche, eds. 1977. *The essential Piaget.* New York: Basic Books.

Grunbaum, A. 1963. *Philosophical problems of space and time.* New York: Knopf (2d ed., Dordrecht: Reidel, 1973).

Habermas, J. 1972. *Knowledge and human interests,* trans. J. Shapiro. Boston: Beacon Press.

Hacking, I. 1975. *The emergence of probability.* Cambridge: Cambridge University Press.

———. 1979. Imre Lakatos's philosophy of science. *British Journal for the Philosophy of Science* 30:181–402.

———. 1983. *Representing and intervening.* Cambridge: Cambridge University Press.

Hallam, A. 1973. *A revolution in the earth sciences.* Cambridge: Cambridge University Press.

Hanson, N. R. 1958. *Patterns of discovery.* Cambridge: Cambridge University Press.

———. 1961. Is there a logic of discovery? In *Current issues in the philosophy of science,* ed. H. Feigl and G. Maxwell, 20–35. New York: Holt, Rinehart & Winston.

Haufling, O. 1981a. *Logical positivism.* Oxford: Basil Blackwell.

———. 1981b. *Essential readings in logical positivism.* Oxford: Basil Blackwell.

Haugeland, J., ed. 1981. *Mind design.* Cambridge: MIT Press.

―――. 1985. *Artificial intelligence: The very idea.* Cambridge: MIT Press.

Heezen, B. C. 1962. The deep-sea floor. In *Continental drift,* ed. S. K. Runcorn, 235–88. New York: Academic Press.

Heirtzler, J. R., and X. Le Pichon. 1965. Crustal structure of the mid-ocean ridges; 3. Magnetic anomalies over the Mid-Atlantic Ridge. *Journal of Geophysical Research* 70:4013–33.

Heirtzler, J. R., X. Le Pichon, and J. G. Baron. 1966. Magnetic anomalies over the Reykjanes Ridge. *Deep-Sea Research* 13:427–43.

Heirtzler, J. R., et al. 1968. Marine magnetic anomalies, geomagnetic field reversals, and motions of the ocean floor and continents. *Journal of Geophysical Research* 73:2119–36. Reprinted in Cox (1973).

Hempel, C. G. 1960. Inductive inconsistencies. *Synthese* 12:439–69.

―――. 1962. Deductive nomological vs. statistical explanation. In *Scientific explanation, space, and time,* ed. H. Feigl and G. Maxwell. Minnesota Studies in the Philosophy of Science, vol. 3. Minneapolis: University of Minnesota Press.

―――. 1965. *Aspects of scientific explanation.* New York: Free Press.

Hempel, C. G., and Paul Oppenheim. 1948. Studies in the logic of explanation. *Philosophy of Science* 15:135–75. Reprinted in Hempel (1965).

Hess, H. H. 1962. History of oean basins. In *Petrologic studies,* ed. A. E. J. Engel, H. L. James, and B. F. Leonard, 599–620. Boulder, Colo.: Geological Society of America. Reprinted in Cox (1973).

Hesse, M. B. 1974. *The structure of scientific inference.* Berkeley: University of California Press.

Hoffmann, G. W. 1985. The role of experiment in the development of the relativistic impulse approximation for proton + nucleus elastic scattering. In *Proceedings of the LAMPF workshop on Dirac approaches to nuclear physics,* ed. J. R. Shepard, C. Y. Cheung, and R. L. Boudrie, 28–72. Los Alamos, N.M.: Los Alamos National Laboratory.

Hoffman, G. W., et al. 1981. Elastic scattering of 500-MeV polarized protons from ^{40,48}Ca, ^{90}Zr, and ^{208}Pb, and breakdown of the impulse approximation at small momentum transfer. *Physical Review Letters* 47:1436–40.

Holland, J. H., K. J. Holyoak, R. E. Nisbett, and P. R. Thagard. 1986. *Induction: Processes of inference, learning, and discovery.* Cambridge: MIT Press.

Hollis, M., and S. Lukes, eds. 1982. *Rationality and relativism.* Cambridge: MIT Press.

Holmes, A. 1929. Radioactivity and earth movements. *Transactions of the Geological Society of Glasgow* 18:559–606.

―――. 1944. *Principles of physical geology.* London: Nelson.

Hooker, C. A. 1978. An evolutionary naturalist realist doctrine of perception and secondary qualities. In *Perception and cognition: Issues in the foundations of psychology,* ed. C. W. Savage. Minnesota Studies in the Philosophy of Science, vol. 9. Minneapolis: University of Minnesota Press.

―――. 1987. *A realistic theory of science.* Albany: State University of New York Press.

Hospers, J. 1951. Remanent magnetism of rocks and the history of the geomagnetic field. *Nature* 168:1111–12.

Howson, C., and A. Franklin. 1985. Newton and Kepler. *Studies in History and Philosophy of Science* 16:379–86.

Hudson, L. 1966. *Contrary imaginations*. New York: Schocken.

Hull, D. 1978. Altruism in science: A sociobiological model of cooperative behavior among scientists. *Animal behavior* 26:685–97.

———. 1982. The naked meme. In *Learning, development, and culture*, ed. H. Plotkin, 273–327. New York: Wiley.

———. Forthcoming. *Science as a process: An evolutionary account of the social and conceptual development of science*. Chicago: University of Chicago Press.

Jeffrey, R. C. 1956. Valuation and acceptance of scientific hypotheses. *Philosophy of Science* 23:237–46.

———. 1965. *The logic of decision*. New York: McGraw-Hill (2d ed., Chicago: University of Chicago Press, 1983).

———. 1983. *Bayesianism with a Human Face*. In *Testing scientific theories*, ed. J. Earman, 133–56. Minnesota Studies in the Philosophy of Science, vol. 10. Minneapolis: University of Minnesota Press.

———. 1985. Probability and the art of judgment. In *Observation, experiment, and hypothesis in modern physical science*, ed. P. Achinstein and O. Hannaway, 95–126. Cambridge: MIT Press.

Jeffreys, H. 1924. *The earth*. Cambridge: Cambridge University Press (5th ed. 1970).

———. 1931. Problems of the earth's crust: A discussion. *Geographical Journal* 78:453.

Johnson-Laird, P. N. 1983. *Mental models*. Cambridge: Harvard University Press.

Kahneman, D., P. Slovic, and A. Tversky, eds. 1982. *Judgment under uncertainty: Heuristics and biases*. Cambridge: Cambridge University Press.

Kasbeer, T. 1973. *Bibliography of continental drift and plate tectonics*. Boulder, Colo.: Geological Society of America.

Kerman, A. K., H. McManus, and R. M. Thaler. 1959. The scattering of fast nucleons from Nuclei. *Annals of Physics* 8:551–635.

Keynes, J. M. 1921. *A treatise on probability*. London: Macmillan.

Kitcher, P. 1981. Explanatory unification. *Philosophy of Science* 48:507–31.

Knorr-Cetina, K. D. 1981. *The manufacture of knowledge*. Oxford: Pergamon Press.

———. 1983. The ethnographic study of scientific work: Towards a constructivist interpretation of science. In *Science observed*, ed. K. D. Knorr-Cetina and M. Mulkay, 115–40. Hollywood, Calif.: Sage.

Knorr-Cetina, K. D., and M. Mulkay, eds. 1983a. Introduction: Emerging principles in social studies of science. In *Science observed*, ed. K. D. Knorr-Cetina and M. Mulkay, 2–17. Hollywood, Calif.: Sage.

———. 1983b. *Science observed*. Hollywood, Calif.: Sage.

Kornblith, H., ed. 1985. *Naturalizing epistemology*. Cambridge: MIT Press.

Kripke, S. 1972. Naming and necessity. In *The semantics of natural language*, ed. G. Harmon and D. Davidson. Dordrecht: Reidel.

Kuhn, T. S. 1961. The function of measurement in modern physical science. *Isis* 52:161–90. Reprinted in Kuhn (1977).

———. 1962. *The structure of scientific revolutions*. Chicago: University of Chicago Press (2d ed. 1970).

———. 1974. Second thoughts on paradigms. In *The structure of scientific theories*, ed. F. Suppe. Urbana: University of Illinois Press. Reprinted in Kuhn (1977).

———. 1977. *The essential tension*. Chicago: University of Chicago Press.

Kyburg, H. E., Jr. 1961. *Probability and the logic of rational belief*. Middletown, Conn.: Wesleyan University Press.

———. 1974. *Logical foundations of statistical inference*. Dordrecht: Reidel.

Lakatos, I. 1970. Falsification and the methodology of scientific research programmes. In *Criticism and the growth of knowledge*, ed. I. Lakatos and A. Musgrave. Cambridge: Cambridge University Press.

———. 1971. History of science and its rational reconstructions. In *Boston studies in the philosophy of science*, vol. 8, ed. R. C. Buck and R. S. Cohen, 91–135. Dordrecht: Reidel.

Langley, P., H. A. Simon, G. L. Bradshaw, and J. M. Zytkow. 1987. *Scientific discovery*. Cambridge: MIT Press.

Lankford, J. 1981. Amateurs versus professionals: The controversy over telescope size in late Victorian science. *Isis* 72:11–28.

Larkin, J., et al. 1980. Expert and novice performance in solving physics problems. *Science* 208:1335–42.

Latour, B., and S. Woolgar. 1979. *Laboratory life*. Beverly Hills, Calif.: Sage.

Laudan, L. 1977. *Progress and its problems*. Berkeley: University of California Press.

———. 1981a. A confutation of convergent realism. *Philosophy of Science* 48:19–48.

———. 1981b. The pseudo-science of science? *Philosophy of the social sciences* 11:173–198. Reprinted in Brown (1984).

———. 1984a. *Science and values*. Berkeley: University of California Press.

———. 1984b. Explaining the success of science: Beyond epistemic realism and relativism. In *Science and reality*, ed. J. T. Cushing, C. F. Delaney, and G. M. Gutting, 83–105. Notre Dame, Ind.: University of Notre Dame Press.

———. 1986. Some problems facing intuitionist meta-methodologies. *Synthese* 67:115–29.

———. 1987. Progress or rationality? The prospects for normative naturalism. *American Philosophical Quarterly* 24:19–31.

Laudan, L., et al. 1986. Scientific change: Philosophical models and historical research. *Synthese* 69:141–223.

Laudan, R. 1980. The method of multiple working hypotheses and the development of plate tectonic theory. In *Scientific discovery: Case studies*, ed. T. Nickles, 331–43. Dordrecht: Reidel.

———. 1981. The recent revolution in geology and Kuhn's theory of scientific change. In *PSA 1978*, vol. 2, ed. P. D. Asquith and I. Hacking, 227–39. East Lansing, Mich.: Philosophy of Science Association.

Laudan, R., and Laudan, L. Forthcoming. Dominance and the disunity of method: Solving the problems of innovation and consensus. *Philosophy of Science*.

Laymon, R. 1980. Idealization, explanation, and confirmation. In *PSA 1980*, vol. 1, ed. P. D. Asquith and R. N. Giere, 336–50. East Lansing, Mich.: Philosophy of Science Association.

———. 1982. Scientific realism and the counterfactual path from data to theory. In *PSA 1982*, vol. 1, ed. P. D. Asquith and T. Nickles, 107–21. East Lansing, Mich.: Philosophy of Science Association.

Leplin, J., ed. 1984. *Scientific realism*. Berkeley: University of California Press.

Levi, I. 1967. *Gambling with truth*. New York: Knopf.

———. 1980. *The enterprise of knowledge*. Cambridge: MIT Press.

Lloyd, E. A. 1984. A semantic approach to the structure of population genetics. *Philosophy of Science* 50:112–29.

Lorenz, K. 1941. Kant's Lehre vom Apriorischen im Lichte gegenwartiger Biologie. *Blaetter für Deutsche Philosophie* 15:94–125.

———. 1962. Kant's doctrine of the a priori in the light of contemporary biology. *General Systems* 7:23–35. Reprinted in Plotkin (1982).

Luce, R. D., and H. Raiffa. 1957. *Games and decisions*. New York: Wiley.

Lugg, A. 1986. An alternative to the traditional model? Laudan on disagreement and consensus in science. *Philosophy of Science* 53:419–24.

Lumsden, C., and E. O. Wilson. 1981. *Genes, mind, and culture*. Cambridge: Harvard University Press.

Lyell, C. 1830. *The principles of geology*. London: John Murray.

McClelland, D. C. 1970. On the dynamics of creative physical scientists. In *The ecology of human intelligence*. Harmondsworth, England: Penguin.

McCloskey, M. 1983. Naive theories of motion. In *Mental models*, ed. D. Gentner and A. L. Stevens, 299–324. Hillsdale, N.J.: Erlbaum.

McCloskey, M., A. Caramazza, and B. Green. 1980. Curvilinear motion in the absence of external forces: Naive beliefs about the motion of objects. *Science* 210:1139–41.

MacKenzie, D. A. 1978. Statistical theory and social interests: A case study. *Social Studies of Science* 8:35–83.

———. 1981. *Statistics in Britain: 1865–1930*. Edinburgh: Edinburgh University Press.

McKinsey, J. C. C., A. C. Sugar, and P. Suppes. 1953. Axiomatic foundations of classical particle mechanics. *Journal of Rational Mechanics and Analysis* 2:253–72.

Mahoney, M. J. 1976. *Scientist as subject: The psychological imperative*. Cambridge, Mass.: Ballinger.

Mannheim, K. 1952. *Essays on the sociology of knowledge*. London: Routledge & Kegan Paul.

March, J. G., and H. A. Simon. 1958. *Organizations*. New York: Wiley.

Marion, J. B. 1970. *Classical dynamics*, 2d ed. New York: Academic Press.

Marvin, U. B. 1973. *Continental drift: The evolution of a concept.* Washington, D.C.: Smithsonian Institution Press.

Merton, R. K. 1938. Science, technology, and society in seventeenth-century England. In *Osiris: Studies on the history and philosophy of science.* Bruges: Saint Catherine Press (2d ed., New York: Harper & Row, 1970).

———. 1949. *Social theory and social structure.* New York: Free Press (rev. eds., 1957, 1968).

———. 1973. *The sociology of science,* ed. N. Storer. New York: Free Press.

Miller, L. D. 1972. Possible validity of the relativistic Hartree-Fock approximation in nuclear physics. *Physical Review Letters* 28:1281–84.

Miller, L. D., and A. E. S. Green. 1972. Relativistic self-consistent Meson field theory of spherical nuclei. *Physical Review C* 5:241–52.

Minsky, M. 1975. A framework for representing knowledge. In *The psychology of computer vision,* ed. P. H. Winston. New York: McGraw-Hill.

Mitroff, I. 1974. *The subjective side of science.* New York: Elsevier.

Montague, R. 1974. Deterministic theories. In *Formal philosophy: Selected papers of Richard Montague,* ed. R. Thomason, 303–59. New Haven: Yale University Press.

Morris, R. G. M. 1983. An attempt to dissociate "spatial-mapping" and "working memory" theories of hippocampal function. In *Neurobiology of the hippocampus,* ed. W. Seifert, 405–32. London: Academic Press.

Moulines, C. U. 1976. Approximate application of empirical theories: A general explication. *Erkenntnis* 10:201–27.

Moulines, C. U., and J. D. Sneed. 1979. Suppes' philosophy of physics. In *Patrick Suppes,* ed. R. J. Bogdan, 59–91. Dordrecht: Reidel.

Mulkay, M. J. 1974. Methodology in the sociology of science: Some reflections on the study of radio astronomy. *Social Science Information* 13:107–19.

———. 1979. *Science and the sociology of knowledge.* London: Allen & Unwin.

———. 1980. Interpretation and the use of rules. In *Science and social structure,* ed. T. Gieryn, 111–25. Transactions of the New York Academy of Sciences, ser. 2, vol. 39.

———. 1981. Action and belief or scientific discourse? A possible way of ending intellectual vassalage in social studies of science. *Philosophy of the Social Sciences* 11:163–71.

Mulkay, M., J. Potter, and S. Yearley. 1983. Why an analysis of scientific discourse is needed. In *Science observed,* ed. K. D. Knorr-Cetina and M. Mulkay, 171–203. Hollywood, Calif.: Sage.

Nagel, E. 1961. *The structure of science.* New York: Harcourt, Brace, and World.

Neisser, U. 1976. *Cognition and reality.* New York: Freeman.

Nersessian, N. J. 1984. *Faraday to Einstein: Constructing meaning in scientific theories.* Dordrecht: Nijhoff.

Neurath, O., R. Carnap, and C. Morris, eds. 1955, 1970. *Foundations of the unity of science.* 2 vols. Chicago: University of Chicago Press.

Newell, A., and H. A. Simon. 1972. *Human problem solving.* Englewood Cliffs, N.J.: Prentice-Hall.

Newell, H. E., et al. 1979. *Impact of technology on geophysics.* Washington, D.C.: National Academy of Sciences.

Newton-Smith, W. H. 1981. *The rationality of science.* London: Routledge & Kegan Paul.

Nickles, T. 1986. Remarks on the use of history as evidence. *Synthese* 69:253–66.

Nickles T., ed. 1980. *Scientific discovery, logic, and rationality.* Dordrecht: Reidel.

Nisbett, R., and L. Ross. 1980. *Human inference: Strategies and shortcomings of social judgment.* Englewood Cliffs, N.J.: Prentice-Hall.

O'Keefe, J. 1983. Spatial memory within and without the hippocampal system. In *Neurobiology of the hippocampus,* ed. J. Seifert, 375–403. London: Academic Press.

O'Keefe, J., and L. Nadel. 1978. *The hippocampus as a cognitive map.* Oxford: Clarendon Press.

Opdyke, N. D., B. P. Glass, J. D. Hays, and J. H. Foster. 1966. Paleomagnetic study of Antarctic deep-sea cores. *Science* 154:349–57.

Pellionisz, A., and R. Llinas. 1982. Space-time representation in the brain: The cerebellum as a predictive space-time metric tensor. *Neuroscience* 7:2949–70.

Phillips, L., and W. Edwards. 1966. Conservatism in a simple probability learning task. *Journal of Experimental Psychology* 72:346–54.

Pickering, A. 1981. The role of interests in high-energy physics: The choice between charm and colour. In *The social process of scientific investigation,* ed. K. D. Knorr, R. Krohn, and R. Whitley. Sociology of the Sciences Yearbook, vol. 4. Dordrecht: Reidel.

———. 1984. *Constructing quarks.* Chicago: University of Chicago Press.

Picklesimer, A. 1985. Features of relativistic and nonrelativistic approaches to proton-nucleus scattering. In *Proceedings of the LAMPF workshop on Dirac approaches to nuclear physics,* ed. J. R. Shepard, C. Y. Cheung, and R. L. Boudrie, 128–59. Los Alamos, N.M.: Los Alamos National Laboratory.

Pitman, W. C., III, and J. P. Heirtzler. 1966. Magnetic anomalies over the Pacific-Antarctic Ridge. *Science* 154:1164–71.

Pitz, G. F., and N. J. Sachs. 1984. Judgment and decision: Theory and application. *Annual Review of Psychology* 35:119–63.

Plotkin, H. C., ed. 1982. *Learning, development, and culture.* New York: Wiley.

Popper, K. R. 1934. *Logic der Forschung: Zur Erkenntnistheorie der modernen Naturwissenschaft.* Vienna: Springer.

———. 1959a. *The logic of scientific discovery.* London: Hutchinson.

———. 1959b. The propensity interpretation of probability. *British Journal for the Philosophy of Science* 10:25–42.

———. 1972. *Objective knowledge.* Oxford: Oxford University Press (2d ed., 1979).

———. 1974. Intellectual autobiography. In *The philosophy of Karl Popper,* 2 vols., ed. P. A. Schilpp. La Salle, Ill.: Open Court.

———. 1983. *Realism and the aim of science.* London: Rowman and Littlefield.

Price, D. J. 1965. Is technology historically independent of science? A study in statistical historiography. *Technology and Culture* 6:553–68.

———. 1984. Notes towards a philosophy of the science/technology interaction. In *The nature of technological knowledge,* ed. R. Laudan, 105–14. Dordrecht: Reidel.

Putnam, H. 1962. What theories are not. In *Logic, methodology, and philosophy of science,* ed. E. Nagel, P. Suppes, and A. Tarski. Palo Alto: Stanford University Press.

———. 1975. The meaning of meaning. In *Minnesota studies in the philosophy of science,* vol. 7, ed. K. Gunderson. Minneapolis: University of Minnesota Press.

———. 1978. *Meaning and the moral sciences.* London: Routledge & Kegan Paul.

———. 1981. *Reason, truth, and history.* Cambridge: Cambridge University Press.

———. 1982. Why reason can't be naturalized. *Synthese* 52:3–23.

———. 1983. *Realism and reason,* vol. 3 of *Philosophical papers.* Cambridge: Cambridge University Press.

Pylyshyn, Z. 1984. *Computation and cognition.* Cambridge: MIT Press.

Quine, W. V. O. 1953. *From a logical point of view.* Cambridge: Harvard University Press.

———. 1969. Epistemology naturalized. In *Ontological relativity and other essays.* New York: Columbia University Press. Reprinted in Kornblith (1985).

Raff, A. D., and R. G. Mason. 1961. Magnetic survey off the West Coast of North America, 40° N latitude to 52° N latitude. *Bulletin of the Geological Society of America* 72:1267–70.

Rahbar, A., et al. 1981. First measurement of the spin rotation parameter Q for p-^{40}Ca elastic scattering at 500 MeV. *Physical Review Letters* 47:1811–14.

Ray, L. 1979. Neutron isotopic density differences deduced from 0.8 GeV polarized proton elastic scattering. *Physical Review C* 19:1855–72.

———. 1983. First order interpretation of optical potentials. In *The interaction between medium energy nucleons in nuclei—1982,* ed. H. O. Meyer, 121–42. New York: American Institute of Physics.

Reichenbach, H. 1928. *Philosophie der Raum-Zeit-Lehre.* Berlin: W. de Gruyter.

———. 1938. *Experience and prediction.* Chicago: University of Chicago Press.

———. 1949. *The theory of probability.* Berkeley: University of California Press.

———. 1958. *The philosophy of space and time,* trans. M. Reichenbach and J. Freund. New York: Dover.

Rensch, B. 1971. *Biophilosophy.* New York: Columbia University Press.

Roe, A. 1952. *The making of a scientist.* New York: Dodd & Mead.

———. 1961. The psychology of the scientist. *Science* 134:456–59.

Rorty, R. 1979. *Philosophy and the mirror of nature.* Princeton: Princeton University Press.

Rosenkrantz, R. D. 1977. *Inference, method, and decision.* Dordrecht: Reidel.

———. 1980. Induction as information acquisition. In *Applications of inductive logic,* ed. L. J. Cohen and M. B. Hesse, 68–89. Oxford: Oxford University Press.

Rubinstein, R. A., C. D. Laughlin, Jr., and J. McManus. 1984. *Science as cognitive process.* Philadelphia: University of Pennsylvania Press.

Runcorn, S. K. 1962. Paleomagnetic evidence for continental drift and its geophysical cause. In *Continental drift,* ed. S. K. Runcorn, 1–65. New York: Academic Press.

Ruse, M. 1981. What kind of revolution occurred in geology? In *PSA 1978,* vol. 2, ed. P. D. Asquith, and I. Hacking, 240–73. East Lansing, Mich.: Philosophy of Science Association.

———. 1986. *Taking Darwin seriously.* Dordrecht: Reidel.

Russell, B. 1912–13. On the notion of cause. *Proceedings of the Aristotelian Society.* Reprinted in Russell (1918).

———. 1914. *Our knowledge of the external world.* London: Allen & Unwin.

———. 1918. *Mysticism and logic.* New York: Longman.

———. 1919. *Introduction to mathematical philosophy.* London: Allen & Unwin.

Salmon, W. C. 1967. *The foundations of scientific inference.* Pittsburgh: University of Pittsburgh Press.

———. 1975. Theoretical explanation. In *Explanation,* ed. S. Körner, 118–45. Oxford: Basil Blackwell.

———. ed. 1979a. *Hans Reichenbach: Logical empiricist.* Dordrecht: Reidel.

———. 1979b. Propensities: A discussion review. *Erkenntnis* 14:183–216.

———. 1984. *Scientific explanation and the causal structure of the world.* Princeton: Princeton University Press.

Savage, L. J. 1954. *Foundations of statistics.* New York: Wiley.

Scheffler, I. 1967. *Science and subjectivity.* New York: Bobbs-Merrill.

Schlosser, R. P. 1984. The 1926 symposium by the American Association of Petroleum Geologists and its impact on the history of the continental displacement theory—A Reappraisal, master's dissertation. Bloomington: Indiana University.

Scriven, M. 1958. Definitions, explanations, and theories. In *Concepts, theories, and the mind-body problem,* ed. H. Feigl, M. Scriven, and G. Maxwell. Minnesota Studies in the Philosophy of Science, vol. 2. Minneapolis: University of Minnesota Press.

———. 1961. The key property of physical laws—Inaccuracy. In *Current issues in the philosophy of science,* ed. H. Feigl and G. Maxwell, 91–101. New York: Holt, Rinehart & Winston.

———. 1962. Explanations, predictions, and laws. In *Scientific explanation, space, and time,* ed. H. Feigl and G. Maxwell. Minnesota Studies in the Philosophy of Science, vol. 3. Minneapolis: University of Minnesota Press.

Sellars, W. 1961. The language of theories. In *Current issues in the philosophy of science,* ed. H. Feigl and G. Maxwell, 57–77. New York: Holt, Rinehart & Winston.

Serot, B. D., and J. D. Walecka. 1986. The relativistic nuclear many-body problem. In *Advances in nuclear physics*, ed. J. W. Negele and E. Vogt. New York: Plenum.

Shank, R. C., and R. P. Abelson. 1977. *Scripts, plans, goals, and understanding.* Hillsdale, N.J.: Erlbaum.

Shapere, D. 1964. The structure of scientific revolutions. *Philosophical Review* 73:383–94.

———. 1974. Scientific theories and their domains. In *The structure of scientific theories*, ed. F. Suppe. Urbana: University of Illinois Press.

———. 1982. The concept of observation in science and philosophy. *Philosophy of Science* 49:485–525.

———. 1984. *Reason and the search for knowledge.* Dordrecht: Reidel.

———. 1985. Objectivity, rationality, and scientific change. In *PSA 1984*, vol. 2, ed. P. Asquith and P. Kitcher, 637–63. East Lansing, Mich.: Philosophy of Science Association.

Shapin, S. 1975. Phrenological knowledge and the social structure of early nineteenth-century Edinburgh. *Annals of Science* 32:219–43.

———. 1979. The politics of observation: Cerebral anatomy and social interests in the Edinburgh phrenology disputes. In *On the margins of science: The social construction of rejected knowledge*, ed. R. Wallis. Sociological Review Monograph no. 27. Keele: University of Keele.

———. 1982. History of science and its sociological reconstructions. *History of Science* 20:157–211.

Shepard, J. R., C. Y. Cheung, and R. L. Boudrie, eds. 1985. *Proceedings of the LAMPF workshop on Dirac approaches to nuclear physics.* Los Alamos, N.M.: Los Alamos National Laboratory.

Shepard, J. R., J. A. McNeil, and S. J. Wallace. 1983. Relativistic impulse approximation for p-Nucleus elastic scattering. *Physical Review Letters* 50: 1443–46.

Shimony, A. 1970. Scientific inference. In *The nature and function of scientific theories*, ed. R. G. Colodny, 79–172. Pittsburgh, Pa.: University of Pittsburgh Press.

———. 1971. Perception from an evolutionary point of view. *Journal of Philosophy* 67:571–83.

———. 1981. Integral epistemology. In *Scientific inquiry and the social sciences*, ed. M. B. Brewer and B. E. Collins, 98–123. San Francisco: Jossey-Bass.

Siegel, H. 1980. Justification, discovery, and the naturalizing of epistemology. *Philosophy of Science* 47:297–321.

———. 1985. What is the question concerning the rationality of science? *Philosophy of Science* 52:517–37.

Simon, H. A. 1945. *Administrative behavior.* New York: Free Press.

———. 1947. The axioms of Newtonian mechanics. *Philosophical Magazine* 36:888–905.

———. 1954. The axiomatization of classical mechanics. *Philosophy of Science* 21:340–43.

————. 1955. Prediction and hindsight as confirmatory evidence. *Philosophy of Science* 22:227–30. Reprinted in Simon (1977).

————. 1957. *Models of man*. New York: Wiley.

————. 1977. *Models of discovery*. Dordrecht: Reidel.

————. 1979. *Models of thought*. New Haven, Conn.: Yale University Press.

————. 1983. *Models of bounded rationality*. 2 vols. Cambridge: MIT Press.

Slater, J. C., and N. H. Frank. 1947. *Mechanics*. New York: McGraw-Hill.

Sneed, J. D. 1971. *The logical structure of mathematical physics*. Dordrecht: Reidel.

Sposito, G. 1976. *An introduction to classical dynamics*. New York: Wiley.

Stegmueller, W. 1976. *The structure and dynamics of theories*. New York: Springer.

————. 1979. *The structuralist view of theories*. Berlin: Springer.

Stewart, J. A. 1986. Drifting continents and colliding interests: A quantitative application of the interests perspective. *Social Studies of Science* 16:261–79.

Stroud, B. 1981. The significance of naturalized epistemology. In *Midwest studies in philosophy*, vol. 6. Minneapolis: University of Minnesota Press. Reprinted in Kornblith (1985).

Suppe, F. 1972. What's wrong with the received view on the structure of scientific theories? *Philosophy of Science* 39:1–19.

————. 1973. Theories, their formulations, and the operational imperative. *Synthese* 25:129–64.

————. ed. 1974. *The structure of scientific theories*. Urbana: University of Illinois Press (2d ed., 1977).

Suppes, P. 1967. What is a scientific theory? In *Philosophy of science today*, ed. S. Morgenbesser, 55–67. New York: Basic Books.

————. 1969. *Studies in the methodology and foundations of science: Selected papers from 1951 to 1969*. Dordrecht: Reidel.

Symon, K. R. 1953. *Mechanics*. New York: Addison-Wesley.

Synge, J. L., and B. A. Griffith. 1959. *Principles of mechanics*, 3d ed. New York: McGraw-Hill.

Talwani, H. S., X. Le Pinchon, and J. R. Heirtzler. 1965. East Pacific Rise: The magnetic pattern and the fracture zones. *Science* 150:1109–15.

Thompson, P. 1983. The structure of evolutionary theory: A semantic approach. *Studies in History and Philosophy of Science* 14:215–29.

Tolman, E. C. 1948. Cognitive maps in rats and men. *Psychological Review* 55:189–208.

Toulmin, S. 1961. *Foresight and understanding*. New York: Harper & Row.

————. 1972. *Human knowledge*. Princeton: Princeton University Press.

Tweney, R. D. 1985. Faraday's discovery of induction: A cognitive approach. In *Faraday rediscovered*, ed. D. Gooding and F. James, 189–209. New York: Stockton.

Tweney, R. D., M. E. Doherty, and C. R. Mynatt, eds. 1981. *On scientific thinking*. New York: Columbia University Press.

Tweney, R. D., and S. A. Yachanin. 1985. Can scientists assess conditional inferences? *Social Studies of Science* 15:155–73.

Umbgrove, J. H. F. 1947. *The pulse of the earth*. The Hague: Nijhoff.

Vacquier, V. 1965. Transcurrent faulting in the ocean floor. In *Symposium on continental drift*, ed. P. M. S. Blackett, E. C. Bullard, and S. K. Runcorn. *Proceedings of the Royal Society of London*, ser. A, 258:77–81.

van der Gracht, W. A. J. M. van Waterschoot, et al. 1928. *Theory of continental drift*. Tulsa, Okla.: American Association of Petroleum Geologists.

van Fraassen, B. C. 1970. On the extension of Beth's semantics of physical theories. *Philosophy of Science* 37:325–39.

———. 1972. A formal approach to the philosophy of science. In *Paradigms and paradoxes*, ed. R. G. Colodny, 303–66. University of Pittsburgh Series in the Philosophy of Science, vol. 5. Pittsburgh: University of Pittsburgh Press.

———. 1980. *The scientific image*. Oxford: Oxford University Press.

———. 1981. Theory construction and experiment: An empiricist view. In *PSA 1980*, vol. 2. ed. P. D. Asquith and R. Giere, 663–78. East Lansing, Mich.: Philosophy of Science Association.

———. 1982. The Charybdis of realism: Epistemological implications of Bell's inequality. *Synthese* 52:25–38.

———. 1985. Empiricism in the philosophy of science. In *Images of science*, ed. P. M. Churchland and C. A. Hooker, 245–308. Chicago: University of Chicago Press.

Vening Meinesz, F. A. 1930. Maritime gravity surveys in the Netherlands East Indies: Tentative interpretation of the results. *Ned. Akad. Wetensch., Proc.*, ser. B, 33:566–77.

Vine, F. J. 1966. Spreading of the ocean floor: New evidence. *Science* 154:1405–15.

Vine, F. J., and D. H. Matthews. 1963. Magnetic anomalies over oceanic ridges. *Nature* 199:947–49.

Vine, F. J., and J. T. Wilson. 1965. Magnetic anomalies over a young oceanic ridge off Vancouver Island. *Science* 150:485–89.

von Neumann, J., and O. Morgenstern. 1944. *Theory of games and economic behavior*. Princeton, N.J.: Princeton University Press.

Walecka, J. D. 1974. A theory of highly condensed matter. *Annals of Physics* 83:491–529.

Wallace, A., and S. K. Fenster. 1969. *Mechanics*. New York: Holt, Rinehart & Winston.

Wason, P. C. 1966. Reasoning. In *New horizons in psychology*, vol. 1, ed. B. Foss. Harmondsworth, England: Penguin.

———. 1977. Self-contradictions. In *Thinking*, ed. P. N. Johnson-Laird and P. C. Wason, 114–28. Cambridge: Cambridge University Press.

Wegener, A. 1911. *Thermodynamik der Atmosphaere*. Leipzig: J. A. Barth.

———. 1915. *Die Enstehung der Kontinente und Ozeane*. Braunschweig: F. Vieweg (2d ed. 1920, 3d ed. 1922, 4th ed. 1929).

————. 1924. *The origin of continents and oceans,* trans. J. G. A. Skerl from the 3d German ed. London: Methuen.

————. 1966. *The origin of continents and oceans,* trans. J. Biram from the 4th German ed. New York: Dover.

Whitehead, A. N., and B. Russell. 1910–13. *Principia mathematica,* 3 vols. Cambridge: Cambridge University Press (2d ed. 1925–27).

Willis, B. 1910. Principles of paleogeography. *Science* 31:241–60.

Wilson, B. R., ed. 1970. *Rationality.* New York: Harper & Row.

Wilson, D. 1983. *Rutherford.* Cambridge: MIT Press.

Wilson, J. T. 1965. A new class of faults and their bearing on continental drift. *Nature* 207:343–47.

————. 1968. Static or mobile earth—The current scientific revolution. *Proceedings of the American Philosophical Society* 112:309–20.

Wimsatt, W. C. 1980. Reductionistic research strategies and their bases in the units of selection controversy. In *Scientific discovery: Case studies,* ed. T. Nickles, 213–59. Dordrecht: Reidel.

Winch, P. 1958. *The idea of a social science.* London: Routledge & Kegan Paul.

Winkler, R. L. 1972. *Introduction to Bayesian inference and decision.* New York: Holt, Rinehart & Winston.

Wittgenstein, L. 1953. *Philosophical investigations,* trans. G. Anscombe. Oxford: Blackwell.

Woolgar, S. 1981. Interests and explanation in the social study of science. *Social Studies of Science* 11:365–94.

————. 1983. Irony in the social study of science. In *Science observed,* ed. K. D. Knorr-Cetina and M. Mulkay, 239–66. Hollywood, Calif.: Sage.

Wuketits, F. M. 1984. *Concepts and approaches in evolutionary epistemology.* Dordrecht: Reidel.

Yearley, S. 1982. The relationship between epistemological and sociological cognitive interests: Some ambiguities underlying the use of interest theory in the study of scientific knowledge. *Studies in History and Philosophy of Science* 13:353–88.

Index

Abraham, R., 87
Achinstein, P., 104, 284 n. 11
Ackermann, R., 288 n. 9
Ad hoc, 223, 224
Adie, R., 253
Approximation, 45–46, 71; puzzles about, 106
Arationality assumption, 41–42, 52
Arbib, M., 281 n. 2
Aristotle, and universal statements, 102
Azande, 4, 56

Barnes, B., 50, 54, 107
Bartlett, F. C., 282 n. 6
Base-rate fallacy, 154–55; in medical diagnosis, 153–54
Bayesian decision models, 145–48; role of new information in, 147–48; and scientists, 156–57, 213
Bayesian information models, 148–49
Bayes's theorem, 148
Beam swinger, 123–24
Beatty, J., xx, 286 n. 10
Berger, P., 58–59
Berlin, B., 109
Birnbaum, A., xix
Black, M., xix
Blackett, P. M. S., 246, 248
Bloor, D., xx, 50–52, 54, 107, 109, 282 n. 13
Bounded rationality, 157–58; as natural cognitive activity, 160–61
Boyd, R., 8, 46, 140, 170, 290 n. 7

Bradie, M., 282 n. 12
Bradshaw, G. L., 281 n. 2
Bromberger, S., 83
Bruhnes, B., 244
Bucher, W., 228, 261
Bullard, E., 254

Callebaut, W., xx
Campbell, D. C., xx, 8, 10, 222, 249, 281 n. 2, 282 n. 12
Caramazza, A., 175–76
Carnap, R., 23, 24, 172, 283 nn. 1, 4, 6, 290 n. 8; and induction, 26–27
Cartesian circle, 11–12
Cartwright, N. D., 90–91, 170, 285 n. 4
Casscells, W., 153
Catastrophism, 229
Causality: and explanation, 104–5; and mechanical systems, 99–100; and modality, 101
Causal schemata, 155–56, 173
Chamberlin, R. T., 238–39, 291 n. 3
Chamberlin, T. C., 239, 291 n. 3
Chess, problem solving in, 88
Churchland, P. M., xix, 7, 8, 136
Churchland, P. S., xix, 81, 109, 110, 135
Clark, B. C., 192–98, 214–16
Classical mechanics, 63–74
Coffa, J. A., 283 n. 1
Cognition, and interests, 164–65
Cognitive maps, 6, 110, 134–36
Cognitive resources, 213–14, 253–54; and scientific interests, 214–21

Collins, H. M., xx, 4, 55
Common cause, principle of, 97–98
Conjunction rule, violation of, 151–53
Constructivism, 56–59; and empiricism, 132–33; limitations of, 130–33; self-applicability of, 288 n. 7
Context of discovery, 24–25, 32–33
Contingency: in the laboratory, 114; and negotiation, 112–15
Contractionist models of the earth, 228–29; and stabilism, 231
Correspondence rules, 24–25, 74–75
Cox, A., 246, 270
Cox, G., 55
Creer, K., 246–48

Dalrymple, B., 242–46, 269
Decision models: basic structure, 143–45; Bayesian, 145–48; for experimental tests, 165–68; role of values in, 143–44, satisficing, 157–59
DeMey, M., 281 n. 2, 291 n. 6
Descartes, R., 11–12
Dietz, R. S., 292 n. 7
Dirac equation, 183, 192–93
Dirac phenomenology, 192–93; predictive success of, 195; response to, 195–98, 203–8
Directional geomagnetism, 246–48
Doherty, M. E., 281 n. 2
Doppelt, G., 44, 284 n. 19
Dorling, J., 157
Douglas, M., 51
Duhem-Quine thesis, 38; technological fix for, 138–39
Durkheim, E., 51, 58
Du Toit, A. L., 250, 253

Edge, D., 16, 50
Edwards, W., 149–50, 151
Eltanin-19, 264–66, 268, 270
Empiricism, 61; constructive, 46–50, 93–94, 171–72; and constructivism, 132–33; logical, 22–28; limitations of, 128–30; varieties of, 94–95
Empiricist repertoire, 60–61
Epigenetic rules, 14
Ericsson, K. A., 179
Evans-Pritchard, E. E., 56

Evolutionary epistemology, 15
Evolutionary models: components of, 17; and history of science, 18; of scientific development, 17–18, 222, 275–76
Ewing, M., 261, 266, 274
Experimental tests: design and execution of, 208–13; model of, 165–68; probability judgments for, 167–68
Experimentation: and realism, 124–25; Murphy's law for, 169
Explanation: causal, 104–5; and prediction, 28; scientific, 28

Farley, J., 53
Faust, D., 6
Feigl, H., 26
Fenster, S. K., 69
Feyerabend, P. K., 131, 284 nn. 16, 21
Fine, A., xx, 170, 288 n. 3
Foster, C., xx
Foster, J., 267–68
Foundationism, 23–24
Frankel, H., xix, 250, 251, 273, 276, 291 n.1, 292 nn. 8, 9
Franklin, A., 157, 288 n.9
Frege, G., 22
Friedman, M., 23

Gale, G., xix, 7
Galileo, 66
Galison, P., 214, 288 n.9
Gallistel, R., 110
Gamow, G., 14
Gardner, H., 281 n. 3
Garfinkel, H., 51
Gaston, J., 284 n. 15
Geometrical cognition, 133–37
Gieryn, T., xx, 291 n. 7
Gieson, G., 53
Gilbert, G. N., 58–61
Glen, W., 276, 291 n. 1, 292 nn. 8, 9, 293 n. 12
Goldman, A., 282 n. 9
Goldstein, H., 65
Goodman, N., 81, 282 n. 10
Goodness of fit, judged by physicists, 190
Green, A., 191
Green, B., 175–76
Griggs, R. A., 179

Gruber, H., xx, 281n.2, 291n.6
Grunbaum, A., xix, 23

Habermas, J., 51, 52
Hacking, I., 38, 145, 165, 281n.1, 288n.2
Hamiltonian, 71–73, 99
Hanson, N. R., 32, 284n.16
Harmonic oscillator. *See* Linear oscillator
Haufling, O., 283n.1
Haugeland, J., 281n.3
Heezen, B., 268, 274
Heirtzler, J. R., 262–64, 265–66, 270–75
Hempel, C. G., 28, 104, 162
Herron, E., 261, 264
Hertz, H., 76
Hess, H., 251–53, 257, 258, 292n.7
Hesse, M. B., 108, 281n.2
Hilbert, D., 22
Holland, J. H., 281n.2
Hollis, M., 55
Holmes, A., 249–51, 273
Holyoak, K. J., 281n.2
Hooker, C. A., 7, 8
Hook's law, 66, 68–70
Howson, C., 157
Hull, D., xx, 282nn.12, 14
Human problem solving: in chess, 88; in physics, 88–89
Hume, D., 23, 26, 170
Hypotheses, 80–81

Identification, 74–76
Impulse approximation, 182; relativistic, 203–6, 290n.4
Incommensurability, 36–37, 275
Indiana University Cyclotron Facility (IUCF), 113, 287n.1; layout, 116
Induction, problem of, 26–27
Instrumentalism, 26
Interests: and cognition, 164–65; and cognitive resources, 214–21, 240; epistemic and nonepistemic, 161–63, 213, 222, 238–41, 272, 274; realignment of, 274; and sociology of science, 52–54; and support of mobilism, 239–40; and support of stabilism, 238–39, 261–64; and values, 161–62
Interpretation, 74–76
Intuitive logician, 176–77

Intuitive physicist, 175–76
Irving, E., 246–48, 267

Jaramillo event, 246, 269–70
Jeffrey, R. C., 145, 148–49, 172
Jeffreys, H., 228, 236, 249–50
Johnson-Laird, P. N., 282n.6
Juan de Fuca Ridge, 257–61, 262, 264
Judgment, 6–7; conservatism in, 149–50; scientific, 141–78
Justification, context of, 24–25, 32–33

Kahneman, D., 6, 150, 168, 172–74, 179; and base-rate fallacy, 154–55; and causal schemata, 155–56; and violation of conjunction rule, 151–53
Kay, P., 109
Kepler's laws, 103
Keynes, J. M., 26
King, L., 253
Kitcher, P., 104
Knorr-Cetina, K. D., xx, 57–59, 288n.5; on Azande, 112–13
Kripke, S., 36
Kuhn, T. S., xv, 16, 18, 22, 63, 108, 191, 275, 281n.2, 283n.1, 284nn.17, 20, 285n.3, 286n.8; and direct modeling, 71; and exemplars, 34–35, 46, 66, 75; and incommensurability, 36; and naturalism, 32–34; political metaphor, 37, 275; and scientific values, 162–63; and sociology of science, 50; theory of science, 34–38
Kyburg, H. E., 149, 172

Laboratory studies, 56–59, 112
Lakatos, I., 38–40, 191, 199, 241
Langley, P., 281n.2
Lankford, J., 53
Laplace, P., 98; and nebular hypothesis, 228
Larkin, J., 89
Latour, B., xx, 4, 56–59, 112
Laudan, L., xix, 7, 8, 16, 55, 191, 241, 282nn.5, 7, 293n.13; and approximation, 106; and historical argument against realism, 107; and metamethodology, 42–43; and problem-solving effectiveness, 40–41, 250–51, 273; and progress,

Laudan, L. (*continued*)
140; and pursuit, 41, 200; and realism,
45–46; and reticulated model, 44–45;
theory of science, 40–46
Laudan, R., xix, 291 nn. 1, 3, 293 n. 13
Laws: of motion, 65–68, 76–78; of scien-
tific development, 16–17; as universal
generalizations, 102–4
Laymon, R., xix, 285 n. 4, 291 n. 5
Le Pichon, X., 262–64, 270–74, 275
Leplin, J., 7
Levi, I., 149, 162
Linear oscillator, 68–74; damped, 73–74;
Hamiltonian formulation for, 71–72
Llinas, R., 135
Lloyd, E., 286 n. 10
Logical empiricism, 22–28
Longwell, C. R., 235–36, 237
Lorentz, K., 282 n. 11
Los Alamos Meson Physics Facility
(LAMPF), 185–90
Luckmann, T., 58–59
Lugg, A., 44
Lukes, S., 55
Lumsden, C., 14
Lyell, C., 229

McCloskey, M., 175–76
Mach, E., 23, 65, 76
MacKenzie, D., 4, 50, 53, 56
McMullin, E., xix
McNeil, J., 203–8
Mahoney, M. J., 281 n. 2
Matthews, D., 253–54
Matuyama, M., 244–45
Mannheim, K., 51, 58
March, J. G., 157
Marsden, J. E., 87
Marx, K., 34, 58
Medical diagnosis, 153–54
Merton, R. K., 29–32, 34, 52, 199,
284 n. 13
Metamethodology: and Lakatos, 40; and
Laudan, 42–43
Mill, J. S., 23, 199
Miller, D., 191
Minsky, M., 282 n. 6
Mitroff, I., 281 n. 2, 293 n. 12
Mobilism, 231

Modality: and causality, 101; and realism,
98–99
Models, 47–48, 78–80; and approaches,
223; and definitions, 79, 82; qualitative
versus quantitative, 273; and reality, 82;
and theories, 82–86
Morley, L., 292 n. 8
Morris, R. G. M., 134–35
Moulines, C. U., 285 n. 4
Mulkay, M., 16, 59–61, 284 n. 14, 291 n. 6
Murnane, K., 285 n. 22
Murphy's law for experimentation, 167
Mynatt, C. R., 281 n. 2

Nadel, L., 110, 135
Naturalism, 7–9; evolutionary, 12–16; and
Kuhn, 32–34; and philosophy of science,
8–12; possibility of, 12; and sociology of
science, 51–52
Nebular hypothesis, 228
Necessity: and causality, 99–100; and pro-
pensity, 101
Negotiation: and contingency, 112–15; and
experimentation, 114–15
Neisser, U., 4, 282 n. 6
Nersessian, N. J., 281 n. 1
Neutrons: detectors for, 122–23; and pro-
tons, 121–22
Newell, A., 88, 179
Newton, I., and motion in resisting media,
73–74
Newton's laws of motion, 65–68, 82,
87–88, 90; as definitions, 76, 87; and in-
tuitive physics, 175–76; law of universal
gravitation, 66, 102–4
Newton-Smith, W. H., 3, 41, 106, 285 n. 23
Nickles, T., xix, 281 n. 1, 283 n. 5
Nietzsche, F., 58
Nisbett, R., 6, 156, 173, 281 n. 2
Norms of science, 29–31, 199, 276–77
Nuclear potential, 180–84; nuclear optical
potential, 181–82

Observation, versus theory, 26, 48–49,
94–95, 128
Oceanography, 242
O'Keefe, J., 110, 135
Olduvai event, 246, 261, 270
Opdyke, N. D., 261, 264, 265, 270, 274

Operational definitions, 74
Oppenheim, P., 28
Overmann, R., xx

Paleomagnetism, 242–46
Pascal, B., 145
Peano, G., 22
Peirce, C. S., 199, 284 n. 10, 286 n. 4
Pellionisz, A., 135
Pendulum, 70–71, 76–78
Phillips, L., 149–50
Piaget, J., 34, 281 n. 2, 282 n. 6
Pickering, A., 131, 281 n. 4, 291 nn. 6, 8
Picklesimer, A., 200–201
Pinch, T. J., xx, 4, 55
Pitman, W., 264–65, 270, 274–75
(*p*, *n*) experiments, 120–21; geometry of, 133
Popper, K. R., 8, 24, 36, 222, 283 nn. 1, 5, 284 n. 8; and induction, 27; and Lakatos, 38–39; and propensities, 286 n.4; and simplicity, 224
Potter, J., 59
Prediction: role in decision making, 201–3, 293 n. 11; what counts as, 199–200; why success matters, 198–203
Preus, J. S., xx
Price, D., 288 n. 9
Priority disputes, 31
Probability: base rate, 154–55; and Bayesian decision models, 145–49; conjunction rule, 151–53; and experimental tests, 167–68, 201–3; role in science, 171–78
Propensity: and necessity, 101–2; and probability, 101–2
Protons: and beam swinger, 123–24; control of beam, 119–20; in elastic scattering, 185–90; and neutrons, 121–22; in (*p*, *n*) reaction, 120–21; polarized, 119, 185–90; production of beam, 115–18; properties of, 121–22; as research tool, 126–27, 129; use in experiments, 120–24
Psychologism, 24–25
Putnam, H., 6, 9, 36, 46, 98, 110, 284 n. 16
Pylyshyn, Z., 281 n. 3

Quine, W. V. O., 9, 38; and Duhem-Quine thesis, 138–39; objections to modality, 100

Ransdell, S. E., 179
Rationality: and biology, 14; bounded, 157–58; categorical versus hypothetical, 7, 9; and decision making, 142; instrumental, 10, 160–61; principles of, 3; of science, 3
Realism: Charybdis of, 97–98; constructive, 50, 92–110; convergent, 45–46; and experimentation, 124–27; historical argument against, 107; in laboratory, 111–40; metaphysical, 6, 98; modal, 98–99; naturalistic, 7–8; unrestricted, 96–97
Reichenbach, H., 23, 24, 283 nn. 2, 5; and induction, 27
Relativism, 2, 37–38; empirical program of, 54–56
Representation, 6
Representativeness heuristic, 152–53
Rescher, N., xix
Reward structure of science, 31
Rorty, R., 6, 281 n. 1
Rosenkrantz, R., 145, 157
Ross, L., 6, 156, 173
Runcorn, S. K., 246–48, 267
Ruse, M., xx, 14, 291 n. 1
Russell, B., 22, 23, 282 n. 3
Rutherford, E., 126–27, 169, 180–81

Salmon, W. C., xix, 97, 172, 282 n. 8, 284 n.11; and causal explanation, 104–5, 287 n. 7
Satisficing, 157–61; and human agents, 159–60; and realism, 171–72; and scientists, 161–65
Savage, L. J., 149
Scheffler, I., 284 n. 18
Schlosser, R., 240, 291 n. 2
Schroedinger equation, 90, 182–84; equivalence with Dirac equation, 224
Schutz, A., 58
Science policy, 10
Scriven, M., 284 n. 16, 285 n. 4
Seafloor sediments, 266–68
Seafloor spreading, 249–52; Hess's model, 251–52; Holmes's model, 249–51; vindication of, 261–70
Selection task, 176–77
Sellars, W., 284 n. 16

Shank, R., 287n.6
Shapere, D., xix, 140, 284nn.18, 21, 285n.4, 286n.8
Shapin, S., 50, 52–53
Shepard, J., 203–8
Shimony, A., xix, 282n.12, 289n.5
Siegel, H., 10, 282n.10
Similarity: and argument against realism, 107–9; between model and world, 81–82; respects of, 92–94
Simon, H. A., 88, 157–58, 160–61, 179, 281n.2, 290n.3
Simplicity, 224–25
Slovic, P., 6
Smith, D., xx
Social structure of science, 29–31
Sociobiology, 14
Squash problem, 174
Stabilism, 231; geological arguments for, 236–38; geophysical arguments for, 235–36
Stegmueller, W., 184, 285n.1
Stephenson, E. J., xx
Stewart, J., 239, 240, 292n.4, 294n.1
Storer, N., 31
Stroud, B., 282n.9
Suppe, F., xix, 62, 95, 128, 283n.6, 285n.4
Suppes, P., xix, 62
Symon, K. R., 65

Taxicab problem, 154–56, 168, 173–74
Technology: and Duhem-Quine problem, 138–39; as embodied knowledge, 140; in nuclear research, 137–38; role in geology, 248–49; and scientific progress, 140
Teller, P., xx
Textbooks, 62–68; organization of, 64–68; sample table of contents, 67
Thagard, P. R., 281n.2
Theories: falsifiable, 223; as formal systems, 25–26, 47, 87–88; general characterization of, 82–86; and models, 62–91; of science, 1, 22–61; semantic view of, 47–48; well defined, 86, 88
Thompson, P., 286n.10
Tolman, E. C., 110, 282n.6

Toulmin, S., 27, 275, 282n.12, 284nn.16, 21
Transform faults, 258
Truth, 79; correspondence theory of, 82; redundancy theory of, 82
Tversky, A., 6, 168, 172–74, 179; and base-rate fallacy, 154–55; and causal schemata, 155–56; and violation of conjunction rule, 151–53
Tweney, R. B., xx, 179, 281n.2

Uniformitarianism, 229

Values: and interests, 161–63; role in decision making, 143–44, 201
Van der Gracht, W. A. J. M. van Waterschoot, 234–35, 240
Van Fraassen, B. C., xix, 7, 46–50, 62, 80, 97–98, 171–72, 288n.4; and approximation, 106; and empirical adequacy, 48–50, 128–30, 193–94; and explanation, 104–5; and observable, 128–30; and phenomenology of scientific activity, 130
Vine, F. J., 252–54, 257–61, 268–70, 274–75, 292n.9
Vine-Matthews hypothesis, 254–57, 292n.8; objections to, 262–64, 270–72
Visualizability, and quantum theory, 136–37

Walecka, J. D., 191–92
Wallace, A., 69
Wallace, S. J., 203–8
Wason, P. C., 176–77, 179
Waters, C. K., xx
Wegener, A., 229–30; cognitive resources, 230; critics, 234–41; decision problem, 231; evidence, 232–34; Greenland experience, 230–31; interests, 231–32
Whewell, W., 199
Whitehead, A. N., 23
Whitmer, J., 292n.5
Willis, B., 228–29, 239
Wilson, E. O., 14
Wilson, J. T., 257–61
Wimsatt, W., 289n.4

Winch, P., 51
Winkler, R., 145
Wittgenstein, L., 51
Woodcock, J., xx
Woodward, J., xix
Woolgar, S., xx, 4, 56–59, 112, 285 n. 26

Worrall, J., xix

Yachanin, S. A., 179
Yearley, S., 59, 285 n. 26

Zytkow, J. M., 281 n. 2